IET

ELECTRICAL INSTALLATION DESIGN GUIDE

Calculations for Electricians and Designers

2nd Edition

Published by The Institution of Engineering and Technology, London, United Kingdom

The Institution of Engineering and Technology is registered as a Charity in England & Wales (no. 211014) and Scotland (no. SC038698).

The Institution of Engineering and Technology is the new institution formed by the joining together of the IEE (The Institution of Electrical Engineers) and the IIE (The Institution of Incorporated Engineers). The new Institution is the inheritor of the IEE brand and all its products and services, such as this one, which we hope you will find useful.

© 2008, 2013 The Institution of Engineering and Technology

First published 2008 (978-0-86341-555-0)
Reprinted September 2008
Reprinted (with amendments) November 2008
Second edition (incorporating BS 7671:2008 (2011)) 2013 (978-1-84919-657-4)
Reprinted June 2014

Copies of this publication may be obtained from:
The Institution of Engineering and Technology
PO Box 96, Stevenage, SG1 2SD, UK
Tel: +44 (0)1438 767328
Email: sales@theiet.org
www.theiet.org/wiringbooks

ISBN 978-1-84919-657-4

Typeset in the UK by Carnegie Book Production, Lancaster
First printed in the UK by Alpine Press Ltd, Kings Langley
Reprinted in the UK by Advent Print Group, Andover

Contents

Cooperating organisations

The IET acknowledges the contribution made by the following organisations in the preparation of this guide.

BEAMA Ltd
P.D. Galbraith IEng MIIE MIEE MCMI

BEAMA Installations Ltd
P. Sayer IEng MIIE GCGI

British Cables Association
C.K. Reed IEng MIIE

Electrical Contractors' Association of Scotland t/a SELECT
D. Millar IEng MIIE MILE

Electrical Safety Council
Eur Ing J. Bradley BSc CEng MIEE FCIBSE

ERA Technology Ltd
M.W. Coates BEng

Institution of Engineering and Technology
P.E. Donnachie BSc CEng FIET
G.G. Willard DipEE CEng MIEE

Society of Electrical and Mechanical Engineers serving Local Government
C.J. Tanswell CEng MIEE MCIBSE

Author
P.R.L. Cook CEng FIEE

Preface

This book provides step-by-step guidance on the design of electrical installations, from domestic installation final circuit design to fault level calculations for LV/large LV systems. Apprentices and trainees will find it very helpful in carrying out the calculations necessary for a basic installation.

The book has also been prepared to provide a design sequence, calculations and data for a complete design. All necessary cable and equipment data to carry out the calculations is included. Consultants will be able to check the calculations of their design packages. It includes calculations and necessary reference data not found in the design packages, such as cable conductor and sheath temperatures and allowances for harmonics.

Design sequence

1

- **Load characteristics**
- **Supply characteristics**
- **Installation outline**
- **Distribution design**

- **Standard final circuits**
- **Isolation and switching**
- **Final assessment**
- **Compliance with BS 7671**

▼ **Figure 1.1** Design sequence

> **1** Determine load characteristics, load centres, maximum demands of load centres, standby system requirements
> **2** Determine supply characteristics, agree tariffs with electricity company
> **3** Prepare installation outline of the distribution system
> **4** Design distribution system
>   **4a** Select protective devices and cables
>   **4b** Consider voltage drop limitations
>   **4c** Calculate prospective fault currents
>   **4d** Calculate distribution and earth fault loop impedances
>   **4e** Select earthing and other protective conductor sizes
> **5** Size final circuits
> **6** Check isolation and switching requirements
> **7** Final assessment
> **8** Sign electrical installation certificate

132.3   1.1   Load characteristics

1.1.1   Load and location (Chapter 3)

The first step in any installation design is to identify the loads and their characteristics in terms of their physical location, kVA demand, power factor, inrush and starting currents etc. Once the loads have been identified the preferred location for the incoming supply can be selected, typically at the load centre of the installation i.e. close to the larger loads. Locating the incoming supply at any other location will increase the cost of the installation and result in a less satisfactory design. Voltages will vary with load and it may be impracticable at a later date to add further loads without substantial reinforcement. High voltage (HV) distribution systems are to be preferred to low voltage where practicable in terms of load demand and facilities for HV/LV transformers. This will reduce voltage drop, which may be important for large loads and for motors/compressors.

132.4   1.1.2   Standby systems

Assessment of the consequences of loss of supply, perhaps by risk assessment with the customer, will determine the need for standby arrangements. Decisions are preferably made at an early stage, as the decision will affect the distribution design.

1

132.2
313.1 ## 1.2   Supply characteristics

An early approach to the electricity distributor after initial determination of the load and load characteristics is necessary to determine likely supply characteristics including:

- ▶ supply voltage (HV or LV)
- ▶ supply system (TN-S, TN-C-S or TT)
- ▶ location.

A tariff agreement with the intended electricity supplier should also be reached, if possible.

1.3   Installation outline

With the supply position determined, a distribution circuit outline can be prepared with sub-distribution boards at each sub load centre, typically each building or each floor of a multi-storey building.

132.1 ## 1.4   Distribution design

132.8 ### 1.4.1   Protective devices and cables (Chapter 4)

Once the distribution boards have been identified together with their maximum demands, protective devices (overcurrent) and cables (or busbar trunking) can be selected sufficient to carry the loads.

1.4.2   Voltage drop (Chapter 5)

With the initial cable conductor size determined and the load known, voltage drop can be calculated and if necessary cable conductor sizes increased to reduce voltage drop.

Voltage drop is calculated or checked prior to determination of fault levels and earth fault loop impedances, as voltage drop is the most likely limiting factor. If a distribution circuit or final circuit meets the voltage drop requirement it is unlikely not to meet the earth fault loop impedance or shock protection requirements.

132.8 ### 1.4.3   Prospective fault currents (Chapter 6)

With the cable conductors sized for voltage drop and of known length, the fault impedances, earth fault loop impedances and prospective fault currents can be calculated. With the prospective fault currents known (maximum and minimum), either three-phase or single-phase as appropriate, switchgear ratings can be selected and type of device determined such as fuse or circuit-breaker type.

1.4.4   Shock protection (Chapter 7)

BS 7671 requires in the event of a fault to earth that a protective device operates within a prescribed time. With conductor size and length determined, the earth fault loop impedance can be calculated and the appropriate protective device selected. This will generally be carried out as part of 1.4.3 above.

1.4.5   Selection of protective conductors (Chapter 8)

If the protective conductors are of the same current-carrying capacity (equivalent cross-sectional area) as the line conductors, no further checks are required. However, it is common practice within the UK to use reduced section protective conductors, for example, the armouring of cables or flat twin with cpc (circuit protective conductor) cables.

Part 7   1.4.6 Special installations or locations

The nature of any special installations or locations will have to be borne in mind from the start of the design as this may affect the supply requirements (caravan parks, marinas), as well as disconnection times etc.

1.5   Standard final circuits

Once the distribution system has been prepared, the designer can move on to the final circuits. The basic design intent is to use standard final circuits wherever possible to simplify design. Chapter 2 (and Appendix B) summarizes the calculations for small installations and final circuits.

132.15   1.6   Isolation and switching

Little calculation is required in determining the isolation and switching requirements; nevertheless it is a most important aspect of design. Well designed and conveniently located facilities for isolation and emergency switching are essential for the safe use of the installation.

1.7   Final assessment

On completion of the design it will normally be necessary to review again the basic requirements with the client, not only with respect to safe working and adequate functionality, but also possible future growth.

1.8   Compliance with BS 7671

120.3   1.8.1 Departures
133.5

Departures from the detail of Parts 3 to 7 of BS 7671 are to be recorded on the Electrical Installation Certificate; however, the fundamental principles of Part 1 are to be complied with.

The use of new materials or inventions that result in departures from BS 7671 is not prohibited providing the resulting degree of safety is not less than that of compliance. These departures are to be recorded on the Electrical Installation Certificate.

134.2   1.8.2 Electrical installation certificates

The designer will accept formal responsibility for the design by certifying to the best of his/her knowledge and belief that the design is in accordance with BS 7671. The design work is not completed until the certificate has been signed.

1

Calculations for Electricians and Designers

Simple installations and final circuits
2

- ■ Supply characteristics
- ■ Fault rating of switchgear
- ■ Final circuit overcurrent devices and cables
- ■ Final circuit voltage drop limitations

- ■ Fault protection
- ■ Short-circuit current protection
- ■ Protective conductors
- ■ Standard final circuits

2.1 Introduction

This chapter provides a straightforward approach to carrying out simple installation and final circuit designs from readily available design data. It is a stand-alone chapter for persons wishing to go straight into such designs. For a more comprehensive approach to installation design, readers can start at Chapter 1 'Design sequence' followed by Chapter 3, and work their way through the fuller approach.

312 2.2 Supply characteristics
313

2.2.1 General

An installation design must start with the supply characteristics, including the earthing arrangements. The information required is that detailed on the BS 7671 Electrical Installation Certificate, part of which is reproduced in Figure 2.1.

The Electricity Safety, Quality and Continuity Regulations 2002 require the electricity distributor to provide the information for that section of the Installation Certificate. The other details are to be provided by the designer and installer.

▼ **Figure 2.1**   Electrical Installation Certificate (extract)

Address: .... Postcode: .... Tel No: ....

SUPPLY CHARACTERISTICS AND EARTHING ARRANGEMENTS Tick boxes and enter details, as appropriate			
Earthing arrangements	**Number and Type of Live Conductors**	**Nature of Supply Parameters**	**Supply Protective Device Characteristics**
TN-C ☐ TN-S ☐ TN-C-S ☐ TT ☐ IT ☐	a.c. ☐ d.c. ☐ 1-phase, 2-wire ☐ 2-wire ☐ 2-phase, 3-wire ☐ 3-wire ☐	Nominal voltage, U/U_o [(1)] ....V Nominal frequency, f [(1)] ....Hz Prospective fault current, I_{pf} [(2)] ....kA	Type: .... Rated current ....A
Other sources of ☐ supply (to be detailed on attached schedules)	3-phase, 3-wire ☐ other ☐ 3-phase, 4-wire ☐	External loop impedance, Z_e [(2)].........Ω (Note: (1) by enquiry, (2) by enquiry or by measurement)	
PARTICULARS OF INSTALLATION REFERRED TO IN THE CERTIFICATE Tick boxes and enter details, as appropriate			
Means of Earthing	**Maximum Demand**		
Distributor's facility ☐	Maximum demand (load) .... kVA / A (delete as appropriate)		
Installation ☐ earth electrode	**Details of Installation Earth Electrode** (*where applicable*)		
	Type (e.g. rod(s), tape etc)	Location	Electrode resistance to earth .... Ω

312.2 ## 2.2.2 Earthing arrangements

In the UK the most common earthing arrangement is now TN-C-S, or PME.

New supplies will almost always be PME (protective multiple earthed) to provide for a TN-C-S system (see Figure 2.2b).

▼ **Figure 2.2a**
TN-S system

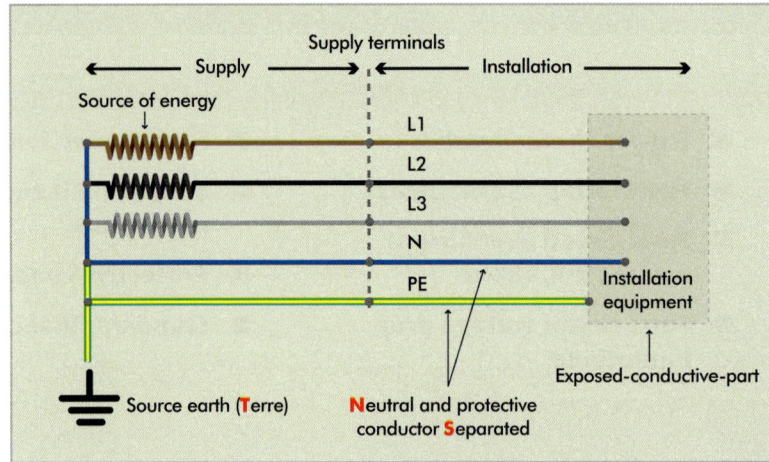

▼ **Figure 2.2b**
TN-C-S system
(PME)

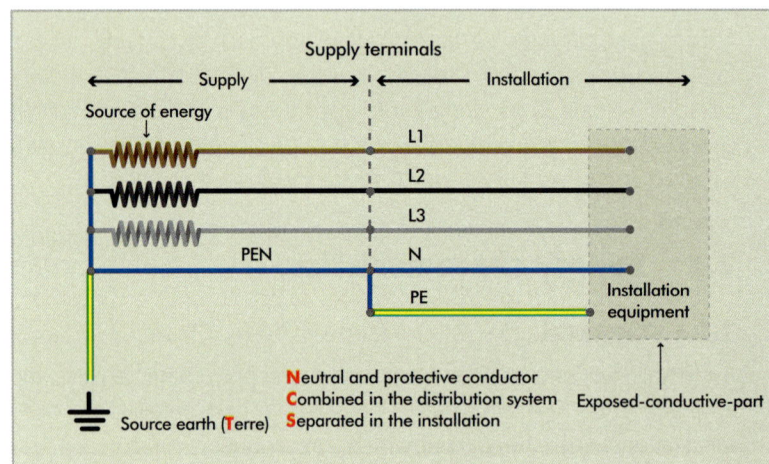

▼ **Figure 2.2c**
TT system

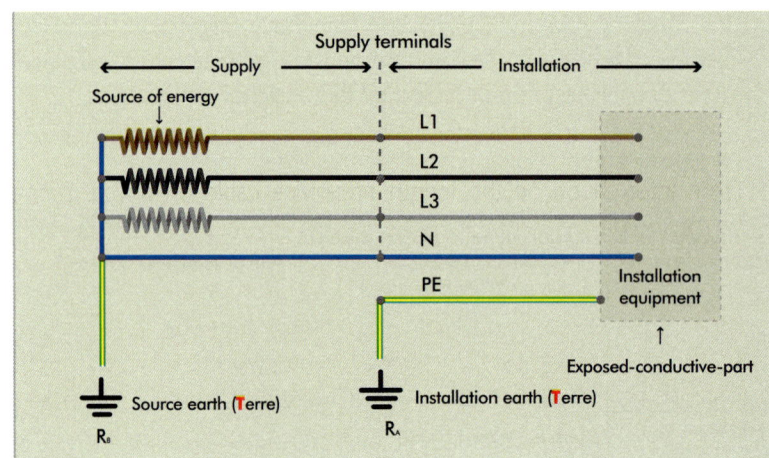

Existing systems may be:

▶ TN-S (supply having a separate earth) or
▶ TN-C-S (supply having a combined neutral and earth) or
312.2 ▶ TT (supply with no earth provided from the source).

(TN-C and IT systems are listed on the form but are not provided by distributors for general application in the UK.)

313   2.2.3   Declared supply characteristics

An enquiry to an electrical distributor for the supply characteristics of a domestic supply will result in the following information being provided:

▶ External earth fault loop impedance Z_e:
    0.35 Ω for PME supplies (TN-C-S systems)
    0.8 Ω for separate earth supplies (TN-S systems)
    21 Ω where no earth is provided (TT systems)
▶ Prospective fault current I_{pf}:
    Excepting certain inner city areas, the maximum prospective fault current I_{pf} will be given as 16 kA.

The lowest external loop impedance given of 0.35 Ω equates to a fault current of only 720 A. A prospective fault current of 16 kA equates to an external loop impedance of approximately 0.015 Ω and is not compatible with a loop impedance of 0.35 Ω. However, these worst-case parameters, if used, enable the designer to have the confidence that the design will be acceptable for any supply, and be valid for the lifetime of the installation even if the supply arrangements outside the dwelling are changed.

BS 7671 allows a design to be carried out on the basis of the actual supply characteristics. However, installation designs generally are carried out before the supply has been provided. Using the supply characteristics above means standard circuits can be developed and used for all dwellings without further calculation. These calculations form the basis of the standard circuits found in such publications as the *Electrician's Guide to the Building Regulations* and the *On-Site Guide*. The formulae for these calculations are given in this chapter and details of the common calculations given in Appendix B.

313.1   2.3   Fault rating of switchgear
512.1.2

All equipment including switchgear assemblies (e.g. comprising busbars, circuit-breakers, controls) must have a fault current rating exceeding the prospective fault current at the point of connection to the installation. Manufacturers will provide assembly ratings, which may be less than the rating of some of the component parts.

From information on the length of service cable on private property, estimates can be made of the attenuation (reduction) in fault level from the figure provided by the distribution company. Chapter 6 provides advice on this. However, for domestic premises consumer units and fuseboards with conditional fault rating of 16 kA can be used, so further calculation is not necessary. Consumer units and fuseboards to 530.3.4   BS EN 60439-3 Annex ZA (Specification for particular requirements of consumer units complete with fuses, c.bs and protective devices) are able to withstand the fault current for prospective fault levels up to 16 kA when the electricity distributor's fuse is a type 2 fuse to BS 88-3 rated at no more than 100 A. The standard 100 A distributor's fuse installed by electricity companies will meet these requirements. Whilst the individual overcurrent devices in the consumer unit or distribution board may not interrupt the fault current at high fault currents, they will be able to carry the currents until the distribution company's fuse operates. Clearly, it is better if, as is usually the case, the fault rating of the fuse or circuit-breaker in the distribution board or consumer unit is of sufficient rating to clear the fault, particularly faults downstream of the consumer unit. Fault levels are rarely as high as 16 kA and rapidly decrease within the installation, so in practice the fuses or circuit-breakers in the consumer unit will clear faults.

Chap 43 ## 2.4 Final circuit overcurrent devices and cables

In this section, the equations to be used are stated at the start of each subsection and then used; full explanation is given in Chapters 4, 5 and 6. However, very simplified explanations are given in each subsection.

2.4.1 Radial final circuits

Derivation of formula

▼ **Figure 2.3**
Coordination of load, device and cable characteristics

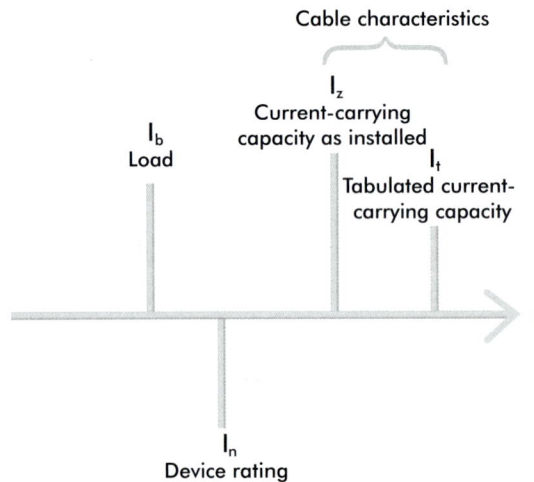

where:

I_b is the design current; it is determined from the load to be connected to the circuit

I_n is the rated current or current setting of the protective device

I_z is the current-carrying capacity of a cable for continuous service in its particular installed condition

I_t is the tabulated current-carrying capacity of a cable found in Appendix F of the *On-Site Guide* or Appendix 4 of BS 7671.

433 **Tabulated cable rating (I_t) and as-installed current-carrying capacity (I_z) formula**

When overcurrent protection is to be provided, the current-carrying capacity of the cable as installed, I_z, is determined by applying rating factors to the tabulated cable ratings, I_t, from Appendix 4 of BS 7671.

Appx 4 $$I_z = I_t\, C_g\, C_a\, C_s\, C_d\, C_i\, C_f\, C_c$$

where:

C_g is the rating factor for grouping, see Table 4C1 of BS 7671 or F3 of the *On-Site Guide*

C_a is the rating factor for ambient temperature, see Table 4B1 of BS 7671 or F1 of the *On-Site Guide*

C_s is the rating factor for thermal resistivity of soil, see Table 4B3 of BS 7671

C_d is the rating factor for depth of buried cable, see Table 4B4 of BS 7671

C_i is the rating factor for conductors surrounded by thermal insulation, see Regulation 523.9 of BS 7671 or Table F2 of the *On-Site Guide*

C_f is the rating factor applied when overload protection is being provided by an overcurrent device with a fusing factor greater than 1.45, e.g. $C_f = 0.725$ for semi-enclosed fuses to BS 3036

C_c is the rating factor for buried circuits: 0.9 for cables buried in the ground requiring overload protection, otherwise is 1.

By inspection of Figure 2.3 it can be seen that the current-carrying capacity of the cable as installed, I_z, must equal or exceed the circuit overcurrent device rated current, I_n.

$$I_z \geq I_n$$

and hence by combining the two equations above, we have

$$I_t \geq \frac{I_n}{C_g C_a C_s C_d C_i C_f C_c} \tag{2.4.1}$$

This equation can be read as, when overcurrent protection is to be provided, the tabulated cable ratings from Appendix 4 of BS 7671 must equal or exceed the circuit overcurrent device rating corrected for the particular installation conditions.

Example 1

A circuit supplying a shower with a loading of 6 kW would have a design current I_b given by:

$$I_b = \frac{6 \times 1000 \text{ W}}{230 \text{ V}} = 26 \text{ A}$$

The nominal current rating in amps, I_n, of the protective device (fuse or circuit-breaker) for a circuit is selected so that I_n is greater than or equal to the design current, I_b, of the circuit.

$$I_n \geq I_b$$

So, in the example of the 6 kW shower I_n must be ≥ 26; select say a 32 A circuit-breaker, that is $I_n = 32$ A. The as-installed cable rating (I_z) must be equal to or greater than 32 A.

$$I_z \geq 32 \text{ A}$$

Notes:
1 Overload protection is provided in practically all circuit designs in order to protect the cable should the load increase above the design value during the life of the installation. However, for a fixed load e.g. a shower circuit, this is not a requirement of BS 7671.
2 The term overcurrent includes both overload current and fault current. Where protection is being provided against overload, protection will also be provided against fault currents. However, the reverse is not true – see sections 4.4 and 4.7 concerning motor circuits.

If overload protection is to be provided apply equation 2.4.1:

$$I_t \geq \frac{I_n}{C_g C_a C_s C_d C_i C_f C_c}$$

Example 2

For the shower circuit above,

> the ambient temperature is assumed (as is usual) to be 30 °C, so C_a is 1
> the cable is not laid in the ground or grouped, so C_g, C_s, C_d and C_c are 1
> the cable is not installed in thermal insulation, so C_i is 1, and
> the circuit protective device is not a semi-enclosed (rewirable) fuse, so C_f is 1.

Hence

$$I_t \geq \frac{32}{1 \times 1 \times 1 \times 1 \times 1 \times 1 \times 1} \geq 32 \text{ A}$$

For a thermoplastic (PVC) insulated and sheathed flat cable with protective conductor from Table 4D5 of BS 7671 or Table F6 of the *On-Site Guide*, installed in an insulated wall, 6 mm² cable is adequate as it has a tabulated rating of 32 A for installation method A.

433.1.103 ## 2.4.2 30 and 32 A ring circuits

Ring circuit cable tabulated rating (I_t) formula

In BS 7671 the rules are changed for the special case of the 30 and 32 A final ring circuit (Regulation 433.1.103), where the overload protection requirements are amended allowing 20 A rated cables ($I_z \geq 20$ A) providing the load current in any part of the ring is unlikely to exceed 20 A.

Therefore $I_z \geq 20$ but $I_z = I_t \, C_g \, C_a \, C_s \, C_d \, C_i \, C_f \, C_c$

Hence

$$I_t \geq \frac{20}{C_g \, C_a \, C_s \, C_d \, C_i \, C_f \, C_c} \tag{2.4.2}$$

Example

A ring final circuit wired with thermoplastic (PVC) insulated cable as per Table 4D5 enclosed in conduit in an insulated wall and protected by a 32 A circuit-breaker ($I_n = 32$ A); the cable rating must equal or exceed 20 A. The cable rating, I_t, is calculated using equation 2.4.2.

If C_g, C_a, C_s, C_d, C_i, C_f and $C_c = 1$ then $I_t \geq 20$ A

From Table 4D5 (column 7) a 2.5 mm² cable is adequate.

2.4.3 Circuits without overload protection

There are many circumstances where the overcurrent device need not provide overload protection, because the load is fixed, as with a shower or luminaire connecting device. It will be providing fault protection. In these circumstances the cable tabulated rating formula for a radial circuit is:

Radial circuit cable tabulated rating (I_t) formula without overload protection

$$I_t \geq \frac{I_b}{C_g\, C_a\, C_s\, C_d\, C_i\, C_f\, C_c} \tag{2.4.3}$$

where I_b is the design current of the circuit.

Note: An adiabatic check will need to be made.

For fuller information see section 4.5.

525,
Appx 4 sect 6

2.5 Final circuit voltage drop limitations

Maximum cable length (L_{vd}) to meet voltage drop limits (i.e. 5% of 230 V = 11.5 V, or 3% for lighting = 6.9 V)

Radial circuits:

For circuits other than lighting

Load at extremity of the circuit

$$L_{vd} = \frac{11.5 \times 1000}{I_b \times (mV/A/m) \times C_t} \tag{2.5.1}$$

For lighting circuits with an evenly distributed load

$$L_{vd} = \frac{6.9 \times 1000}{(I_b/2) \times (mV/A/m) \times C_t} \tag{2.5.2}$$

For ring final circuits

$$L_{vd} = \frac{4 \times 11.5 \times 1000}{I_b \times (mV/A/m) \times C_t} \tag{2.5.3}$$

where:

I_b is design current
(mV/A/m) is the voltage drop per ampere per metre from Appendix 4 of BS 7671
L_{vd} is cable length if the voltage drop limit is not to be exceeded
C_t is a rating factor that can be applied if the load current is significantly less than I_z, the current-carrying capacity of the cable in the particular installation conditions. If C_t is taken as 1, this will err on the safe side. This factor compensates for the temperature of the cable at the reduced current being less than the temperature applicable to the tabulated (mV/A/m) data. See section 5.6.

$$C_t = \frac{230 + t_p - \left(C_g^2 \, C_a^2 \, C_s^2 \, C_d^2 - \dfrac{I_b^2}{I_t^2}\right)(t_p - 30)}{230 + t_p}$$

(2.5.4)

where t_p is the maximum permitted normal operating conductor temperature (°C).

2.5.1 Radial final circuit voltage drop

The voltage drop in an installation must not exceed a value appropriate for the safe and effective functioning of the equipment to be supplied. (See chapter 5 for a fuller treatment). For final circuits, section 6.4 of Appendix 4 of BS 7671 recommends for installations connected directly to a public low voltage network a maximum voltage drop of 3% for lighting and 5% for other uses, that is 6.9 V and 11.5 V respectively at 230 V nominal voltage to Earth.

The current-carrying capacity tables in Appendix 4 include figures for voltage drop per ampere per metre (mV/A/m).

▼ **Figure 2.4**
Radial socket-outlet circuit

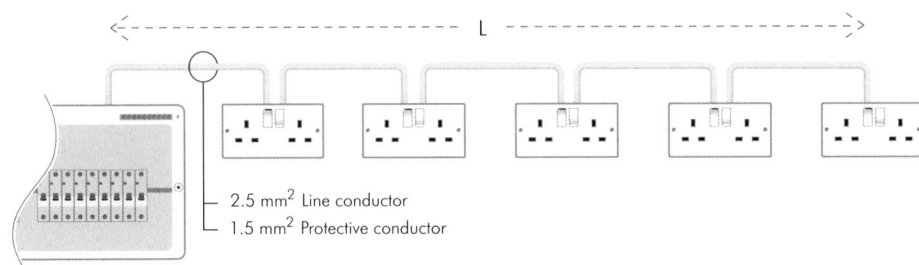

2.5 mm^2 Line conductor
1.5 mm^2 Protective conductor

For loads at the extremity of the circuit, the voltage drop in volts is given by:

Voltage drop = (mV/A/m) x I_b x L x C_t/1000

where L is cable length and I_b is design current.

The maximum length of cable allowed (L_{vd}) if the voltage drop is to be within voltage drop limits is given by:

$$L_{vd} = \frac{\text{max voltage drop} \times 1000}{I_b \times (mV/A/m) \times C_t}$$

Now 5% of 230 V is 11.5 V, so if the voltage drop is not to exceed 5% in a final circuit,

$$L_{vd} = \frac{11.5 \times 1000}{I_b \times (mV/A/m) \times C_t}$$

(2.5.1)

If the load is evenly distributed along the circuit the voltage drop is reduced. For example, consider a lighting circuit with evenly distributed luminaires; the average current is half the total load current (I_b), and the voltage drop is limited to 3% (6.9 V), then:

$$L_{vd} = \frac{6.9 \times 1000}{(I_b/2) \times (mV/A/m) \times C_t}$$

(2.5.2)

2.5.2 Ring final circuit voltage drop

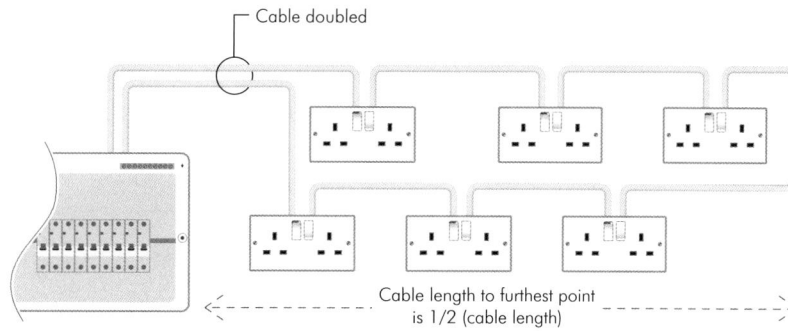

To the furthest point in effect two cables are run, and the length of cable to the furthest point is half the length of cable in the ring, giving a multiplier of 4 for the installed length of cable.

For ring circuits the equation becomes:

$$L_{vd} = \frac{4 \times 11.5 \times 1000}{I_b \times (mV/A/m) \times C_t} \qquad (2.5.3)$$

The current-carrying capacity of the cable of the ring must be not less than 20 A, see Regulation 433.1.103 of BS 7671.

Example

A ring final circuit wired with 2.5 mm^2 cable per Table 4D5, (mV/A/m) = 18 mV/A/m and assuming circuit I_b is 32 A and C_t is 1 then

$$L_{vd} = 4 \times 11.5 \times 1000/32/18 = 79.9 \text{ m}$$

However, in a ring circuit the current in the ring will not be the same all around the ring. If it is assumed to be 20 A at the far end and an additional 12 A is evenly distributed, the average current is (32 + 20)/2 = 26 A and

$$L_{vd} = 4 \times 11.5 \times 1000/26/18 = 98.3 \text{ m}$$

If it is wished to calculate C_t (see section 5.6) and assuming C_a, C_g, C_s and C_d are 1, I_t is 20 A (Reference Method A), cable I_b is 26/2 A, and t_p is 70 °C, then from equation 2.5.4 $C_t = 0.923$ and $L_{vd} = 98.3/0.923 = 106.5$ m. (The average current in each of the two legs of the ring is 26/2 A.)

2.6 Fault protection (protection against indirect contact)

The fault protection requirements can be met by meeting the disconnection times of Regulation 411.3.2. This is achieved by limiting the earth fault loop impedance of the circuits to the values given in Tables 41.2, 41.3 and 41.4 of BS 7671.

For limited disconnection times, the equations to be met are:

for radial circuits: $Z_{41} \geq Z_e + (R''_1 + R''_2)C_r \times L$

for ring circuits: $Z_{41} \geq Z_e + (R''_1 + R''_2)C_r \times \dfrac{L}{4}$

where:

Z_{41} is the maximum earth fault loop impedance given by the appropriate Table 41.2, 41.3 or 41.4 of BS 7671

Z_e is the earth fault loop impedance external to the circuit – in this section it is assumed to be that of the supply, i.e. 0.8 or 0.35 Ω

R''_1 is the resistance per metre of the line conductor (see Table F.1, Table I1 of the *On-Site Guide*)

R''_2 is the resistance per metre of the protective conductor (see Table F.1, Table I1 of the *On-Site Guide*)

C_r is the rating factor for operating temperature (see Table F.3, Table I3 of the *On-Site Guide*)

L is the length of cable in the circuit.

The maximum circuit cable length, L_s, that will limit the circuit resistance such that disconnection in the event of a fault to earth occurs within the required time (0.4 or 5 s) is given by:

for radial circuits: $L_s = \dfrac{Z_{41} - Z_e}{(R''_1 + R''_2)\,C_r}$ (2.6.1)

for ring circuits: $L_s = \dfrac{4(Z_{41} - Z_e)}{(R''_1 + R''_2)\,C_r}$ (2.6.2)

Values of C_r are given in Table F.3 (Table I3 of the *On-Site Guide*). C_r will be taken as 1.20 to correct from 20 °C to a conductor operating temperature of 70 °C, which is appropriate for the thermoplastic (PVC) insulated cables we are using.

For final circuits in domestic premises the installation impedance is taken as the resistance of the line conductor R_1 plus the resistance of the protective conductor R_2. For these circuits the inductance can be neglected, as the cable sizes are less than 25 mm². Conductor resistances are normally given in milliohms per metre at 20 °C, see Table F.1 (Table I1 of the *On-Site Guide*).

To obtain the actual resistances during a fault the tabulated figures R''_1 and R''_2 in mΩ/m must firstly be divided by 1000, then multiplied by the cable length in metres and corrected to working temperature by the factor C_r. The figures in Table F.1 are given at 20 °C, but the conductor operating temperature at full load for thermoplastic cables (PVC) is 70 °C, so a correction C_r must be applied from Table F.3 or *On-Site Guide* Table I3.

Example 1: radial final circuit

Take a 6 kW shower wired in 6/2.5 mm^2 twin with cpc cable with a 32 A type B circuit-breaker from a consumer unit at the origin. Z_{41} from Table 41.3 is 1.44 Ω, assume PME supply $Z_e = 0.35$ Ω, $R''_1 + R''_2$ from Table F.1 is 10.49 mΩ/m or 10.49/1000 Ω/m and C_r from Table F.3 is 1.20. Applying equation 2.6.1 gives

$$L_s = \frac{Z_{41} - Z_e}{(R''_1 + R''_2) C_r} = \frac{1.44 - 0.35}{\left(\dfrac{10.49}{1000}\right) \times 1.20} = 86.5 \text{ m}$$

Example 2: ring final circuit

Take a ring circuit wired in 2.5/1.5 mm^2 twin with cpc cable from a consumer unit at the origin, a PME supply and a 32 A type B circuit-breaker. Z_{41} from Table 41.3 is 1.44 Ω, $Z_e = 0.35$ Ω, $R''_1 + R''_2$ from Table F.1 is 19.51 mΩ/m or 19.51/1000 Ω/m and C_r from Table F.3 is 1.20. Applying equation 2.6.2 gives

$$L_s = \frac{4(Z_{41} - Z_e)}{(R''_1 + R''_2) C_r} = \frac{4(1.44 - 0.35)}{\left(\dfrac{19.51}{1000}\right) \times 1.20} = 186.2 \text{ m}$$

435.1 ## 2.7 Short-circuit current protection

When a circuit is protected by an RCD, the RCD will trip in the event of a fault between a line conductor and the protective conductor. The RCD will not operate in the event of a fault between live conductors, including line to neutral.

If the line to neutral loop impedance exceeds the values given in Tables 41.3 and 41.4 the overcurrent device will not trip in the given times, that is 5 s or instantaneously for circuit-breakers. BS 7671 allows this situation in Regulation 435.1, providing overload protection is given by the overcurrent device and the adiabatic equation of Regulation 434.5.2 is met.

For simplicity, designers will probably wish for operation of the overcurrent device in the times of Tables 41.3 and 41.4, in which case circuit lengths (L_{ss}) need to be limited as follows:

$$L_{ss} = \frac{(Z_{41} - Z_e)}{2R_1 C_r} \quad \text{for radial circuits} \tag{2.7.1}$$

$$L_{ss} = \frac{4(Z_{41} - Z_e)}{2R_1 C_r} \quad \text{for ring circuits} \tag{2.7.2}$$

543.1.3 ## 2.8 Protective conductors

Note: Chapter 8 gives fuller information on protective conductors.

Where an overcurrent device is not providing protection against overload, or where the protective conductor is of a smaller size or lower current-carrying capacity than the line conductor, a check that the adiabatic equation is met must be made:

$$S \geq \frac{\sqrt{I^2 t}}{k}$$

For circuit-breakers see Tables 8.3 and 8.4.

For fuses, the equations which then have to be complied with are:

for radial circuits: $\quad L_a = \frac{Z_a - Z_e}{(R''_1 + R''_2) C_r}$ (2.8.1)

for ring circuits: $\quad L_a = \frac{4(Z_a - Z_e)}{(R''_1 + R''_2) C_r}$ (2.8.2)

where:

L_a is the maximum cable length if the limitation of the adiabatic equation is to be met

Z_a is the maximum adiabatic loop impedance, see Chapter 8 for explanation and tables of values

Z_e is the external or supply loop impedance

R''_1 is the resistance per metre of the line conductor

R''_2 is the resistance per metre of the protective conductor

C_r is a multiplier to convert conductor resistance at 20 °C to the resistance at conductor operating temperature (see Table F.3).

Please note that equation 2.8.2 does not make allowances for spurs from ring circuits and both equations assume a fault at the extremity of the circuit.

2.9 Standard final circuits

Calculations for the standard circuits in the *Electrician's Guide to the Building Regulations* are included in Appendix B. The calculations are presented in a tabular form using the equations of this chapter.

Maximum demand and diversity

3

- ■ **Introduction**
- ■ **Installation outline**
- ■ **Final circuit current demand**
- ■ **Diversity between final circuits**
- ■ **Complex installations**

3.1 Introduction

132.3
311.1

The demand of a circuit or of an installation is the current taken by the circuit or installation over a period of time, say 30 minutes.

Some loads such as a tungsten lamp (light bulb) make a constant demand, say for a 100 W lamp (at 230 V, 100 W/230 V = 0.43 A), 0.43 A all the time it is switched on.

A load such as a washing machine has a number of components to the load perhaps independently controlled. A washing machine may have:

a a variable speed motor for wash, rinse, low speed spin and high speed spin
b a pump to extract water at certain times in the wash cycles
c a 2.5 kW water heater controlled by a programmer and thermostat.

The load of such equipment will vary over time, see Figure 3.1.

Even in a simple installation such as a house there are many loads. Not all the loads are on at the same time. The loads are diverse in:

a current demand
b control mechanisms
c time of use.

This load diversity results in the total load rarely if ever equalling the sum of the individual loads. It would be a wasteful design if advantage were not taken of this diversity between loads.

Cables and other equipment maximum capability are limited by the temperature rise caused by the load current. There is a time element in this; a cable will take an hour to come up to temperature. Short-time high current demands may be acceptable.

This diversity results in complicated demand patterns, see Figure 3.2.

Calculations for Electricians and Designers | **25**
© The Institution of Engineering and Technology

▼ **Figure 3.1**   Kitchen appliance demands (amperes)

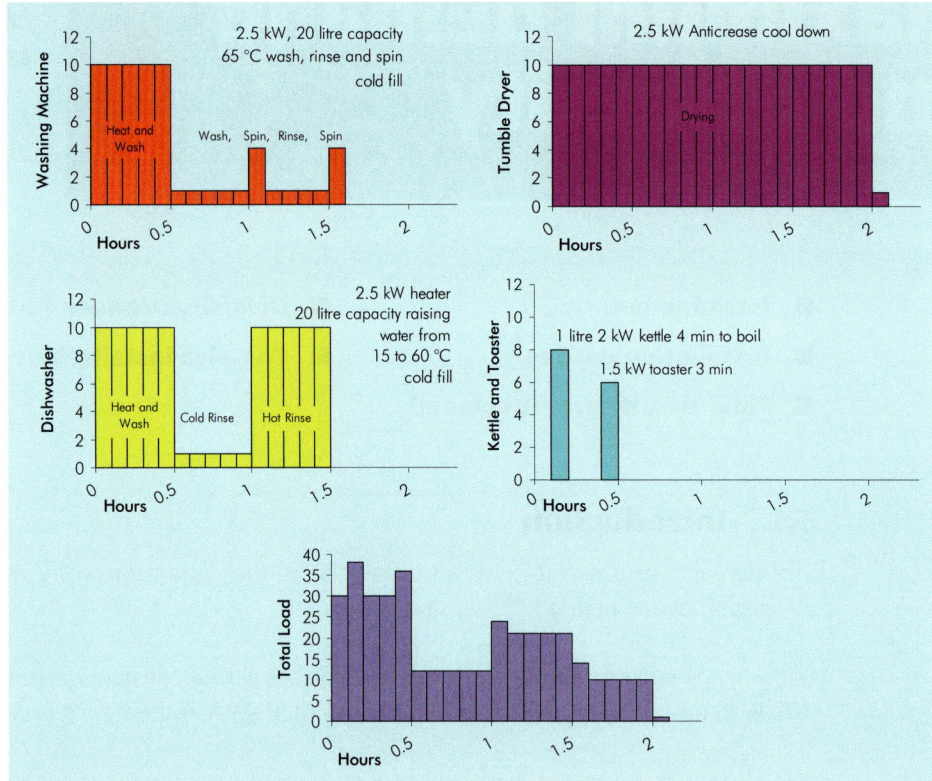

▼ **Figure 3.2**   Half hour domestic demands (kW)

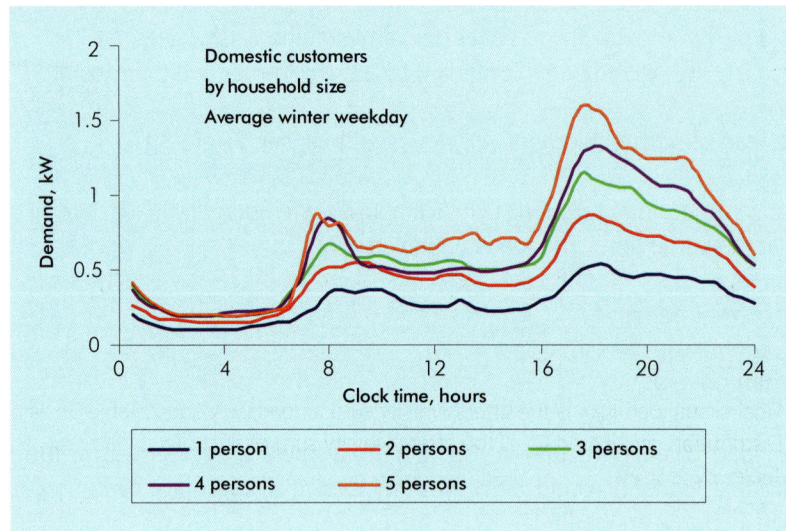

This chapter explains how allowances can be made for diversity between loads in sizing final circuits and distribution circuits. The first step is to identify the loads and their characteristics. This guide includes tables to assist in calculating demands so that cables and switchgear can be selected.

314 ## 3.2 Installation outline

The first step in preparing an installation design is to identify all the electrical loads. The physical position of each load is required as well as the kVA demand, power factor, voltage, frequency etc.

▼ **Figure 3.3**
Installation outline

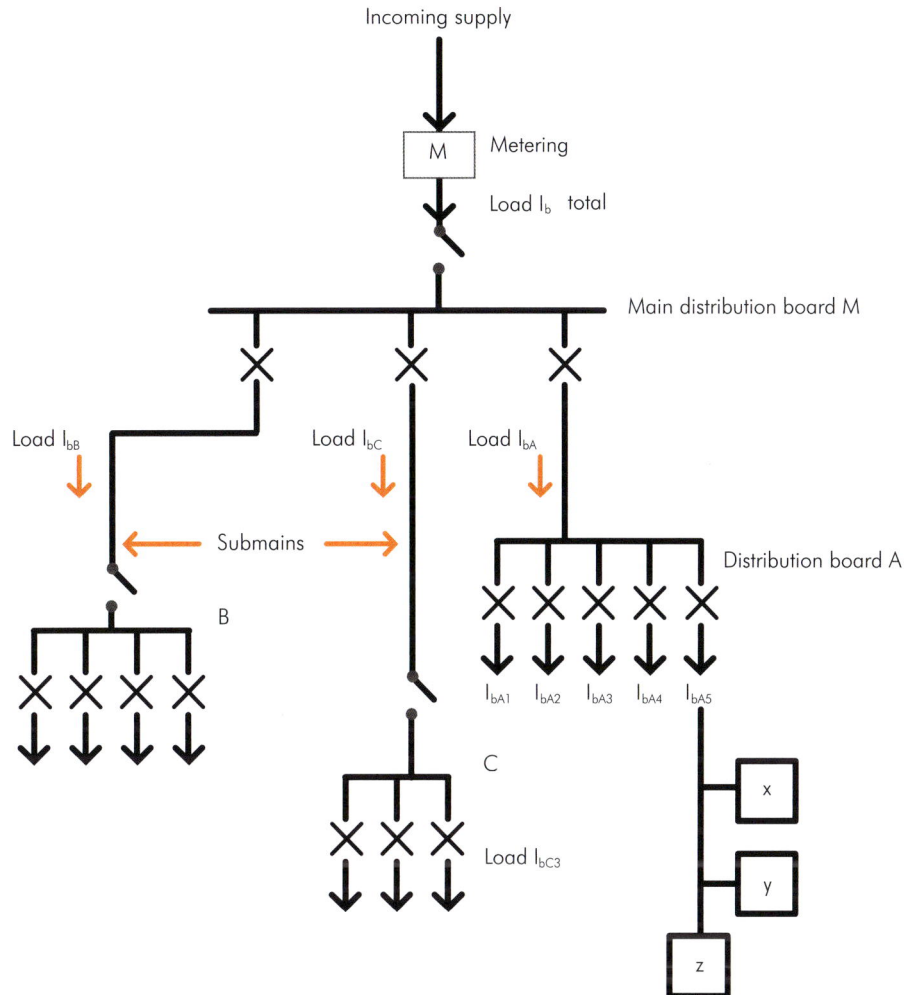

Notes:
1 Final circuit load I_{bA5} is the after-diversity sum of loads x, y and z. Table 3.1 is used.
2 Distribution circuit load I_{bA} is the after-diversity sum of final circuit loads I_{bA1} to I_{bA5}. Table 3.2 is used.
3 Installation load I_b total is the after-diversity sum of distribution circuit loads I_{bA}, I_{bB} and I_{bC}. No table is provided, diversity depends upon load characteristics. This total is the maximum demand of the installation.

An installation outline can then be drawn (Figure 3.3). With the position of the main incoming supply, main distribution board, submains and sub-distribution boards decided, final circuits can be outlined and demand on each element (final circuits, sub-distribution boards, submains, main distribution board) estimated. The incoming supply should be as close to the load centre as is practicable.

The basic design intent is to use standard final circuits wherever possible to avoid repeated design. Provided earth fault loop impedances are below 0.35 or 0.8 ohm at the final distribution board, the standard circuits can be used as the basis of all final circuits. Chapter 2 and Appendix B summarize the calculations for small installations and final circuits.

3.3 Final circuit current demand

The current demand of a final circuit is estimated by adding the current demands of all points of utilization (e.g. socket-outlets) and items of equipment connected to the circuit and where appropriate making allowances for diversity. Table 3.1 (Table A1 of the *On-Site Guide*) gives current demands to be used for final circuits.

▼ **Table 3.1**   Final circuit current demand to be assumed for points of utilization and current-using equipment

	Point of utilisation or current-using equipment	Current demand to be assumed
1	Socket-outlets other than 2 A socket-outlets and other than 13 A socket-outlets[1]	Rated current
2	2 A socket-outlets	At least 0.5 A
3	Lighting outlet[2]	Current equivalent to the connected load, with a minimum of 100 W per lampholder
4	Electric clock, shaver supply unit (complying with BS EN 61558-2-5), shaver socket-outlet (complying with BS 4573), bell transformer, and current-using equipment of a rating not greater than 5 VA.	May be neglected
5	Household cooking appliance	The first 10 A of the rated current plus 30% of the remainder of the rated current plus 5 A if a socket-outlet is incorporated in the control unit
6	All other stationary equipment	British Standard rated current, or normal current

Notes:

1 See Chapter 2 and Appendix B for the design of standard circuits using socket-outlets to BS 1363-2 and BS EN 60309 (BS 4343).

2 Final circuits for discharge lighting must be arranged so as to be capable of carrying the total steady current, viz. that of the lamp(s) and any associated controlgear and also their harmonic currents. Where more exact information is not available, the demand in volt-amperes is taken as the rated lamp watts multiplied by not less than 1.8. This multiplier is based upon the assumption that the circuit is corrected to a power factor of not less than 0.85 lagging, and takes into account controlgear losses and harmonic current.

Because of the change in 1995 in nominal supply voltage (U_0) from 240 V ±6% to 230 V +10% to −6%, calculating load current from kW or kVA rating is made more difficult. Whilst the nominal voltage was officially changed, the actual range stayed much the same and the actual distributed voltage to homes and other premises remained unchanged. Manufacturers may state equipment ratings at 240 V. In determining current demand from kW or kVA demand, the voltage at which the demand is calculated may be used and this may be 240 V and not 230 V.

3.3.1 Examples of circuit current demand

Example 1: shower circuit
See row 6 of Table 3.1

Rating of shower (P) from nameplate 7.2 kW at 240 V

For single-phase circuits, power P in kW $= \dfrac{U_0 I \cos \varnothing}{1000}$

where:

    P   is the power in kW (1000 W)
    I   is current (A)
    U_0 is nominal line voltage (V) at which the load is declared
    $\cos \varnothing$ is power factor (sinusoidal).

Hence current demand of shower circuit $I = \dfrac{1000P}{U_0 \cos \varnothing}$

$\cos \varnothing = 1$ for a resistive load

Hence, $I = \dfrac{1000 \times 7.2}{240 \times 1} = 30 \text{ A}$

Example 2: cooker circuit
See row 5 of Table 3.1

Consider an electric cooker with:

▶   Hob comprising 4 off 3 kW elements
▶   Main oven 2 kW
▶   Grill/top oven 2 kW.

Therefore the total installed capacity = 16 kW at 240 V

As in example 1, $I = \dfrac{1000P}{U_0 \cos \varnothing}$ and $\cos \varnothing = 1$ for a resistive load

Hence, $I = \dfrac{1000 \times 16}{240 \times 1} = 67 \text{ A}$

From Table 3.1, row 5 circuit load is:

'The first 10 A of the rated current plus 30% of the remainder of the rated current plus 5 A if a socket-outlet is incorporated in the control unit.'

Therefore $I = 10 + \dfrac{30(67 - 10)}{100} = 27.1 \text{ A}$

Hence, a 30 or 32 A circuit is selected.

Example 3: lighting circuit

See row 3 of Table 3.1

Consider 10 downstairs lights, assume 100 W demand per lighting point

Therefore circuit demand $I = \dfrac{10 \times 100}{240 \times 1} = 4.17$ A

A 5 A circuit (or 6 A if a BS EN 60898 circuit-breaker is used) would be suitable for tungsten lamps. (Tungsten lamps exhibit a brief inrush current when initially switched on. This inrush current is normally ignored for multiple GLS lamps that are not switched in large groups. For tungsten flood lamps and heat lamps, account should be taken of the inrush current.) However, if extra-low voltage or discharge lighting is to be supplied by a type B circuit-breaker specify a 10 A circuit to reduce unwanted tripping on switch-on due to starting surge. A 6 A type C circuit-breaker may also be appropriate.

For all but the simplest circuits the load characteristics should be assessed and manufacturer's data applied for the selection of all fuses and circuit-breakers. Manufacturer's data should be consulted in particular for tungsten flood lamps and heat lamps, discharge lighting, transformers, motors etc. See Appendix C for an extract from one manufacturer's guidance.

Example 4: immersion heater circuit

See row 6 of Table 3.1

Consider a 3 kW immersion heater

Circuit demand $I = \dfrac{3000}{240} = 12.5$ A

Example 5: motor circuit

See row 6 of Table 3.1

Consider a 3 kW single-phase motor with a power factor of 0.8

Current demand of motor circuit $I = \dfrac{1000P}{U_0 \cos \varnothing} = \dfrac{1000 \times 3}{240 \times 0.8} = 15.6$ A

3.4 Diversity between final circuits

3.4.1 Simple installations

The allowances for diversity in Table 3.2 (Table A2 of the *On-Site Guide*) are for very specific situations and can only provide guidance. The figures given in the table may need to be increased or decreased, depending on the particular circumstances.

The current demand of a group of final circuits supplied from one distribution board (and submain) may be estimated using the allowances for diversity given in Table 3.2. The allowances of rows 1 to 5 are applied to the current demand of items of equipment supplied by the sub-distribution board. The allowances in row 9 (conventional circuits) are applied to the rated current of the overcurrent protective device for the circuit.

The use of other methods of determining maximum demand is not precluded where specified by a competent electrical design engineer.

After the design currents for all the final circuits and submains have been determined, and the conductor sizes chosen, it is necessary to check that the limitations on voltage drop are met.

▼ **Table 3.2**   Allowances for diversity between final circuits for sizing distribution circuits

| Purpose of final circuit fed from conductors or switchgear to which diversity applies | Type of premises | | |
	Individual household installations including individual dwellings of a block	Small shops, stores, offices and business premises	Small hotels, boarding houses, guest houses, etc.
1 Lighting+	66% of total current demand	90% of total current demand	75% of total current demand
2 Heating and power (but see 3 to 8 below)+	100% of total current demand up to 10 amperes +50% of any current demand in excess of 10 amperes	100% f.l. of largest appliance +75% f.l. of remaining appliances	100% f.l. of largest appliance +80% f.l. of second largest appliance +60% f.l. of remaining appliances
3 Cooking appliances	10 amperes +30% f.l. of connected cooking appliances in excess of 10 amperes +5 amperes if socket-outlet incorporated in control unit	100% f.l. of largest appliance +80% f.l. of second largest appliance +60% f.l. of remaining appliances	100% f.l. of largest appliance +80% f.l. of second largest appliance +60% f.l. of remaining appliances
4 Motors (other than lift motors which are subject to special consideration)	not applicable	100% f.l. of largest motor +80% f.l. of second largest motor +60% f.l. of remaining motors	100% f.l. of largest motor +50% f.l. of remaining motors

continues

▼ **Table 3.2** *continued*

Purpose of final circuit fed from conductors or switchgear to which diversity applies		Type of premises		
		Individual household installations including individual dwellings of a block	Small shops, stores, offices and business premises	Small hotels, boarding houses, guest houses, etc.
5	Water-heaters (instantaneous type)*	100% f.l. of largest appliance +100% f.l. of second largest appliance +25% f.l. of remaining appliances	100% f.l. of largest appliance +100% f.l. of second largest appliance +25% f.l. of remaining appliances	100% f.l. of largest appliance +100% f.l. of second largest appliance +25% f.l. of remaining appliances
6	Water-heaters (thermostatically controlled)	no diversity allowable†	no diversity allowable†	no diversity allowable†
7	Floor warming installations	no diversity allowable†	no diversity allowable†	no diversity allowable†
8	Thermal storage space heating installations	no diversity allowable†	no diversity allowable†	no diversity allowable†
9	Standard arrangement of household and similar final circuits (in accordance with Appendix H of *On-Site Guide*)+	100% of current demand of largest circuit +40% of current demand of every other circuit	100% of current demand of largest circuit +50% of current demand of every other circuit	100% of current demand of largest circuit +50% of current demand of every other circuit
10	Socket-outlets other than those included in 9 above and stationary equipment other than those listed above	100% of current demand of largest point of utilization +40% of current demand of every other point of utilization	100% of current demand of largest point of utilization +70% of current demand of every other point of utilization	100% of current demand of largest point of utilization +75% of current demand of every other point in main rooms (dining rooms, etc.) +40% of current demand of every other point of utilization

f.l. = full load

Notes:

* For the purpose of this table, an instantaneous water-heater is deemed to be a water-heater of any loading which heats water only while the tap is turned on and therefore uses electricity intermittently.

† It is important to ensure that the distribution boards and consumer units are of sufficient rating to take the total load connected to them without the application of any diversity.

+ The current demand may be that estimated for example in accordance with Table 3.1. Where the circuit is a standard circuit for household or similar installations the current demand is the rated current of the overcurrent protective device of the circuit.

Example 1: small office

Consider a small office comprising:

i   lighting: 20 off 100 W fluorescent luminaires
ii power: a 10 kW machine and a 2 kW machine (both at 230 volts)
iii socket-outlet circuits: two 32 A circuits

and estimate the maximum demand.

i   lighting: 20 off 100 W fluorescent luminaires

$$\text{lighting demand} = \frac{90}{100} \times \frac{20 \times 100 \times 1.8}{240} = 13.5 \text{ A}$$

Note: See note 2 to Table 3.1 for 1.8 factor, see row 1 column 3 of Table 3.2 for diversity of 90%. If the luminaire manufacturer's VA figures are available, then they should be used instead. Simply divide the VA figure by 240 and multiply by the 90/100 – this generally gives a lower but more accurate figure.

ii   power: a 10 kW machine and a 2 kW machine (both at 230 V)

$$\text{power load} = \left(\frac{100}{100} \times \frac{10 \times 1000}{230}\right) + \left(\frac{75}{100} \times \frac{2 \times 1000}{240}\right) = 43.48 + 6.52 = 50 \text{ A}$$

Note: See row 2 column 3 of Table 3.2 for diversity of 100% for largest, 75% for second machine.

iii   socket-outlet circuits: two 32 A circuits

$$\text{socket-outlet demand} = \left(\frac{100}{100} \times 32\right) + \left(\frac{70}{100} \times 32\right) = 32 + 22.4 = 54.4 \text{ A}$$

Note: See row 10 column 3 of Table 3.2 for diversity of 100% for largest, 70% for second.

Total = 13.5 + 50 + 54.4 = 117.9 A

The designer may know the total demand on the socket circuits is 15 business machines of 100 W rating, e.g. computers, and revise the socket circuit after-diversity demand to

$$\text{Socket after-diversity demand} = \frac{15 \times 100}{230} = 6.25 \text{ A};$$ and revise the after-diversity

demand to

Total = 13.5 + 50 + 6.52 = 70.02 A, and design for 80 A maximum demand.

Example 2: domestic installation

Consider a domestic installation comprising:

i   an electric cooker
ii a 7.2 kW shower
iii two lighting circuits
iv two socket-outlet circuits.

From row 9 of Table 3.2, circuit load (for individual households, including individual dwellings of a block) is:

'100% of current demand of largest circuit + 40% of current demand of every other circuit'.

100% of largest circuit:

cooker as example 2 in 3.3.1   27.1 A

40% other:

7.2 kW shower as example 1 in 3.3.1	30 A	
lighting as 3.3.1 circuit 1	4.17 A	
circuit 2	4.17 A	
socket-outlets circuit 1	30 A	
circuit 2	30 A	
Total other	98.34	
40% other		39.34 A

Installation demand   **66.44 A**

Considering further say 8 dwellings in a block supplied from one submain (e.g. riser). Allowance for diversity is made for the dwellings on any one submain and between submains as necessary, and then between blocks of dwellings. For guidance in such circumstances see Figure 3.4 and 'Complex installations' below.

3.5   Complex installations

The following has been abstracted from an unpublished (Jan 1979) IEC Document 64(Secretariat)254 'Estimation of maximum demand'.

For blocks of residential dwellings, large hotels, industrial and large commercial premises, the allowances are to be assessed by an experienced person.

3.5.1 Accuracy

Estimates of maximum demand can rarely be made accurately. The guidance given here indicates very approximate values with wide tolerances, and must be subject to many reservations. The designer will need to decide:

i   whether they can use values known to them personally or from reliable sources;
ii whether the values given in Tables 3.3 and 3.4 and Figure 3.4 are applicable; and
iii whether the values given in the tables and figure will need to be altered, taking into account:
  a the time profiles of the loads
  b the coincidence of individual loads with other loads – a chart may be helpful in this respect
  c the relationship of the electrical loading of motors to the mechanical load. The mechanical load is a more accurate guide to the electrical load than motor rating – motor ratings are often conservatively selected (overrated)
  d heating and cooling loads, the seasonal demands and how these might coincide with production demands
  e the availability of other sources of supply

f the allowances, if any, for spare capacity or load growth – this must be discussed with the client

g any special considerations which apply to the particular job in hand.

3.5.2 Estimation method

The maximum demand of an installation P_{max} is the sum of the loads installed, P_i, multiplied by a demand factor g:

$$P_{max} = g\, P_i$$

where:

P_i is total installed load for the installation considered, being the sum of all the loads directly connected, generally on the basis of continuous duty

g is demand factor, that is the ratio of the maximum demand of an installation to the corresponding total installed load.

Table 3.3 provides g for some typical complete installations. It is not suitable for intermediate distribution boards supplying mainly or wholly one type of load, e.g. lighting or heating. The factors for particular loads are given in Table 3.4.

Table 3.3 demand factors may be used for sub-distribution boards where there is a typical mix of load, but care needs to be taken; if a particular type of load predominates it will invalidate the use of Table 3.3.

Sub-distribution point estimation

For intermediate distribution boards where one type of load may predominate, Table 3.4 provides demand factors for each type of load. The maximum demand on the distribution board (P_{DB}) is then given by:

$$P_{DB} = P_1 g_1 + P_2 g_2 + P_3 g_3 + P_n g_n$$

where:

P_n is the sum of the particular types of load, e.g. lighting

g_n is a factor to be applied for that load in the particular type of premises.

Tolerance

The difficulty of estimating demand accurately is mentioned in earlier paragraphs. The maximum error in Tables 3.3 and 3.4 is +10%.

It must be remembered that the factors given in Tables 3.3 and 3.4 are purely guidance; particular information relevant to a specific installation will always override this guidance.

▼ **Table 3.3** Demand factors for complete installations

Installation building/premises		Demand factor g for main supply intake	Remarks
1	**Dwellings**		
1.1	Individual	0.4	
1.2	Blocks of flats		The demand factor has to be chosen
1.2.1	without electric heating (as main form of heating) or air-conditioning		from the graphs given in Figure 3.4 according to the mean value of the
1.2.1.1	with lighting and some small appliances only	Figure 3.4 curves A, B and C	loads connected with each flat.
1.2.1.2	fully 'electrified' but without electric heating and air-conditioning	Figure 3.4 curves A, B and C	
1.2.2	with electric heating (as main form of heating) or air-conditioning		The total supply demand will result from the sum of the demand for
1.2.2.1	general demand	Figure 3.4 curves A, B and C	heating and air-conditioning, and all other power demands, see Table 3.4.
1.2.2.2	heating and cooling demand	0.8–1.0	A reduced demand is to be expected when the heating or other loads are controlled so as not to coincide with other applications.
2	**Public buildings**		
2.1	Hotels, boarding houses, furnished apartments	0.6–0.8	
2.2	Small offices	0.5–0.7	
2.3	Large offices (banks, insurance companies, public administration)	0.7–0.8	
2.4	Shops	0.5–0.7	
2.5	Department stores	0.7–0.9	
2.6	Schools	0.6–0.7	
2.7	Hospitals	0.5–0.75	
2.8	Assembly rooms (sports grounds, theatres, restaurants, churches)	0.6–0.8	
2.9	Terminal buildings (railway stations, airports)	requires investigation	
3	**Mechanical engineering industry**		In general the motor drives are over-rated for the mechanical load.
3.1	Metal workings	0.25	
3.2	Car plants	0.25	
4	**Pulp and paper mills**	0.5–0.7	The number of rolling reserve drives considerably affects the demand factor.
5	**Textile industry**		
5.1	Spinning mills	0.75	
5.2	Weaving mills and mixed process installations	0.6–0.7	
6	**Raw material industry**		
6.1	Wood industry	data not available	
6.2	Rubber industry	0.6–0.7	
6.3	Leather industry	data not available	
7	**Chemical industry** Petroleum industry	0.5–0.7	Because the processes in chemical industries are very sensitive to supply failures, the supply must be secure or backup provided. The capacity of a supply is no guarantee of its security, unless I_B is greater than I_n.

continues

▼ Table 3.3 *continued*

Installation building/premises		Demand factor g for main supply intake	Remarks
8	**Cement mills**	0.8–0.9	Reference: Production level about 3500 tons a day with about 500 motors (large mills are driven by HV motors).
9	**Food industry**		
9.1	General (including process engineering)	0.7–0.9	
9.2	Silos	0.8–0.9	
10	**Coal mining**		
10.1	Hard coal preparation	0.8–1.0	
10.2	Underground	1.0	
	Lignite:		
	– general	0.7	
	– excavation	0.8	
11	**Iron and steel mills** (blast-furnaces, converters)		
11.1	Blowers	0.8–0.9	
11.2	Auxiliary drives	0.5	
12	**Rolling mills** (general)	0.5–0.8*	*g depends on the number of standby drives.
12.1	Water supply	0.8–0.9*	
12.2	Ventilation	0.8–0.9*	
	Auxiliaries for rolling mills:		
	– with cooling bed	0.5–0.7*	
	– with loopers	0.6–0.8*	
	– with cooling bed and loopers	0.3–0.5	
12.3	Finishing lines	0.2–0.6+	
13	**Floating docks**		
13.1	Pumping operation during lifting	0.9	Pumping and repair work do not occur simultaneously.
13.2	Repair work without pumping	0.5	
14	**Lighting of street tunnels**	1.0	
15	**Traffic installations**	1.0	Escalators, tunnel (ventilation), traffic lights.
16	**Power generation**		
16.1	Power stations, general	requires investigation	
16.1.1	auxiliary power for low voltage circuits	requires investigation	
16.1.2	emergency supply	1.0	
16.2	Nuclear power stations special power demand, e.g. for trace heating for sodium pipes	1.0	
17	**Cranes**	0.7 per crane	Crane work with intermittent duty: power demand depends on the kind of premises where they are used (e.g. harbours, steel mills, dockyards).
18	**Lifts**	0.5 (highly variable – time of the day)	For simultaneous starting of several cranes or lifts the voltage drop has to be considered.

▼ **Table 3.4** Estimated values for demand factors g for certain loads intended for use in estimating the demands on intermediate and sub-distribution boards

Type of premises	Lighting (1)	Socket-outlets (2)	Water heating not central (3)	Water heating centralised (4)	Cooking, canteens (5)	Refrigeration (6)	Domestic appliances (fixed) (7)	Signalling and address system (8)	Lifts, escalator (9)	Heating and air-conditioning (10)	Air-conditioning not central (11)	Electronic data processing (12)	Experimental and demonstration units (13)	Floodlight installations (14)
1 Individual dwellings	0.6	0.2	0.5	1	0.75	–	0.7	–	0.5	1	0.8		–	–	–
2 Hotels, etc.	0.7	0.1	0.5	1	0.80	0.8	–	0.5	0.5	1	–	requires investigation	–	1
3 Small offices	0.8	0.1	0.3	1	0.50	0.4	–	–	0.7	1	–		–	–
4 Large offices	0.8	0.1	0.3	1	0.80	0.4	–	0.5	0.7	1	–		–	1
5 Shops	0.9	0.3	0.6	1	0.50	0.6	–	–	0.7	1	–		0.2	–
6 Department stores	0.9	0.2	0.3	1	0.80	0.6	–	0.5	0.7	1	–		0.2	1
7 Schools	0.9	0.1	0.3	1	0.80	0.4	–	–	–	1	–		0.4	–
8 Universities and colleges	0.8	0.1	0.3	1	0.80	0.4	–	0.5	0.2	1	–		0.4	–
9 Hospitals	0.7	0.1	0.7	1	0.80	0.8	–	0.5	0.5	1	–		–	–
10 Assembly rooms, public halls	0.9	0.1	0.3	1	0.80	0.6	–	0.5	0.5	1	–		0.4	1

Socket-outlet circuits

The estimation of demand on socket-outlet circuits presents obvious difficulty. The values given in Table 3.4 only apply to circuits comprising a number of outlets which are not expected to be loaded fully and simultaneously. For commercial and industrial installations, a specific estimate of the demand based upon the predicted usage of the sockets needs to be made. For dwellings, the following provides guidance:

Number of circuits	g
1	1.0
2	0.6
4	0.3
8	0.15

Where fixed equipment is fed by socket-outlets, e.g. water heaters or space heating, specific allowance for this must be made. It is to be noted that the g factors above for socket-outlets are for estimation of the effect of the socket-outlets on the maximum demand. Each socket circuit must be designed for its own maximum demand.

▼ **Figure 3.4**
Estimated values for demand factors for the calculation of power demand depending on the number of dwellings (from IEC 64(Sec)254)

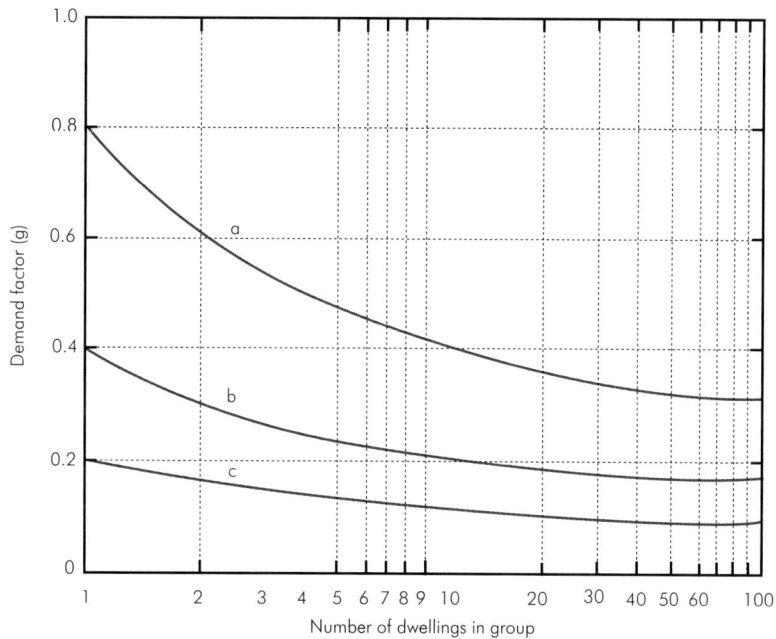

Explanation:

g is the demand factor, i.e. the ratio of the maximum demand of an installation to the installed load. Curve a) high coincidence of usage, b) typical, mixed domestic loads and c) low coincidence.

Example 2 in 3.4.1: domestic installation cont'd

Considering say 8 dwellings in a block.

From curve b for 8 dwellings, demand factor for typical mixed domestic is 0.22.

After-diversity demand for submain: $66.44 + 0.22(8-1)66.44 = 168.76$ A.

Hence, a 200 A riser may be selected.

(Note this is an example of how factors are applied. Actual demands must be discussed and agreed with all parties including the electricity distributor.)

3

40 Calculations for Electricians and Designers
© The Institution of Engineering and Technology

Selection of cables for current-carrying capacity

4

- **Symbols**
- **Preliminary design**
- **Overcurrent requirements**
- **Current-carrying capacity tables**
- **Protection against overload and short-circuit**
- **Protection against fault current only**
- **Corrections for grouping**
- **Motors**

4.0 Symbols

The symbols used in this chapter of the guide are those used in BS 7671 as follows:

I_z	the current-carrying capacity of a cable for continuous service, under the particular installation conditions concerned
I_t	the value of current tabulated in Appendix 4 of BS 7671 for the type of cable, ambient temperature and installation method concerned
I_b	the design current of the circuit, i.e. the current intended to be carried by the circuit in normal service
I_n	the rated current or current setting of the device protecting the circuit against overcurrent
I_2	the operating current (i.e. the fusing current or tripping current for the conventional operating time) of the device protecting the circuit against overload
C	the rating factor to be applied where the installation conditions differ from those for which values of current-carrying capacity are tabulated (in Appendix 4 of BS 7671). The various rating factors are identified as follows:
C_g	the rating factor for grouping
C_a	the rating factor for ambient temperature
C_s	the rating factor for thermal resistivity of the soil
C_d	the rating factor for depth of buried cable
C_i	the rating factor for conductors embedded in thermal insulation
C_f	the rating factor for overcurrent devices where $I_2 \geq 1.45\, I_n$
C_c	the rating factor for circuits buried in the ground
C_t	the rating factor for operating temperature of conductor

4

Appx 4 ## 4.1 Preliminary design

As described in Chapter 3, Figure 3.3, the designer will have prepared a general installation outline and having made due allowances for diversity determined the design current I_b of final circuits, distribution boards, submains, main distribution board etc.

With the design current (I_b) of each circuit determined (see Chapter 3), and the load profile, in particular starting currents, the nominal current rating of the protective devices (I_n) and the cable type and rating for the circuits can be selected.

431.1.1 ## 4.2 Overcurrent requirements

Each installation and every circuit within that installation must be protected against overcurrent by devices which will operate automatically to prevent injury to persons (and livestock) and damage to the installation including the cables. The overcurrent devices must be of adequate breaking capacity and be so constructed that they will interrupt the supply without danger and ideally will facilitate restoration. Cables must be able to carry these overcurrents without damage. Overcurrents may be:

1 fault currents (earth fault or short-circuit) or
2 overload currents.

434 ### 4.2.1 Fault currents

Fault currents arise as a result of a fault in the cables or the equipment. There is a sudden increase in current, perhaps 10 or 20 times the cable rating, the current being limited by the impedances of the source of supply, the cables, the fault itself and the return path. The current is normally of short duration, but fault protection is nonetheless required except in exceptional circumstances (see Regulation 434.3 of BS 7671).

433 ### 4.2.2 Overload currents

Overload currents do not arise as a result of a fault in the cable or equipment. They arise because the current has been increased by the addition of further load.

Overload protection is only required if overloading is possible. It would not be required for a circuit supplying a fixed load.

The load on a circuit supplying a (say) 7.2 kW shower will not increase unless the shower is replaced by one having a higher rating, when the adequacy of the circuit must be checked.

A distribution circuit supplying a number of buildings could be overloaded by additional machinery being installed in one or more of the buildings supplied.

Whereas overload currents are likely to be of the order of 1½ to 2 times the rating of the cable, as mentioned above fault currents may be of the order of 10 to 20 times the rating.

Overloads of less than 1.2 to 1.6 times the device rating, dependent upon the type of device (see Table 4.1 and section 4.2.3 'Small overloads'), are unlikely to result in operation of the device. Regulation 433.1 requires that every circuit shall be designed so that small overloads of long duration are unlikely to occur.

Calculations for Electricians and Designers
© The Institution of Engineering and Technology

It is usual for one device in the circuit to provide both fault protection and any overload protection that is required. The common exception is the overcurrent devices in motor circuits, where the overcurrent device at the origin of the circuit provides protection against fault currents and the motor starter provides protection against overload. Whilst overload protection may not be necessary in a circuit the cable selected must be of a sufficient size to carry the load current.

The current-carrying capacity of cable for continuous service under the particular conditions of its installation (I_z) must be equal to or greater than the current for which the circuit is designed (I_b), that is a current intended to be carried in normal service:

$$I_z \geq I_b \tag{4.2.1}$$

The nominal current or current setting of the protective device (I_n) must also be equal to or greater than the design current of the circuit:

$$I_n \geq I_b \tag{4.2.2}$$

Equations 4.2.1 and 4.2.2 must always be complied with.

(If the load current I_b is say 30 A, the device rating I_n must be equal to or greater than 30 A, say 30 A or 32 A.)

Where overload protection is being provided, the current-carrying capacity of cable for continuous service under the particular conditions of its installation I_z must be equal to or greater than I_n, the nominal current or current setting of the protective device.

$$I_z \geq I_n$$

$$I_z \geq I_n \geq I_b \tag{4.2.3}$$

433.1 4.2.3 Small overloads

The standards for overcurrent devices (fuses and circuit-breakers) specify a current below which the device must not operate (fuse or trip). This is the non-fusing or non-tripping current (I_1), see Table 4.1. They also specify the current at which the device must operate in a specified conventional time. This is the fusing or tripping current (I_2), see Table 4.1.

Overloads below the non-fusing or non-tripping current (I_1) will not operate the overcurrent device.

Overloads between I_1 and I_2 will take longer than the conventional time; for further information see section 6.2.1 of the *Commentary on IEE Wiring Regulations*.

▼ **Table 4.1**   Non-fusing (I_1) and fusing (I_2) currents

Device type	Rated current I_n (A)	Non-fusing* current I_1 or I_{nf} (A)	Fusing* current I_2 or I_f (A)	Conventional fusing* time (hours)
BS EN 60269-2	<16	1.25 I_n for 1 hour	1.6 I_n	1
(BS 88-2, 88-3)	$16 < I_n \leq 63$	1.25 I_n for 1 hour	1.6 I_n	1
	$63 < I_n \leq 160$	1.25 I_n for 2 hours	1.6 I_n	2
	$160 \leq I_n\ 400$	1.25 I_n for 3 hours	1.6 I_n	3
	$400 < I_n$	1.25 I_n for 4 hours	1.6 I_n	4
BS 1361 type 1	$5 < I_n \leq 45$		1.5 I_n	4
type 2	$60 < I_n \leq 100$		1.5 I_n	4
BS 1362	$I_n \leq 13$	1.6 I_n	1.9 I_n	
1, 2, 3, 4 m.c.bs	≤ 10	1.0 I_n for 2 hours	1.5 I_n	1
to BS 3871	> 10	1.0 I_n for 2 hours	1.35 I_n	1
B, C, D circuit-breakers	≤ 63	1.13 I_n	1.45 I_n	1
to BS EN 60898	> 63	1.13 I_n	1.45 I_n	2
BS 3036	5	1.8 I_n	$\leq 2.0\ I_n$	0.75
	15	1.8 I_n	$\leq 2.0\ I_n$	1.00
	30	1.8 I_n	$\leq 2.0\ I_n$	1.25
	60	1.8 I_n	$\leq 2.0\ I_n$	1.5
	100	1.8 I_n	$\leq 2.0\ I_n$	2.0
	150	1.8 I_n	$\leq 2.0\ I_n$	2.0
	200	1.8 I_n	$\leq 2.0\ I_n$	2.5

* 'fusing' for fuses, 'tripping' for circuit-breakers.

Appx 4   ## 4.3   **Current-carrying capacity tables**

4.3.1 Tabulated current-carrying capacity I_t

Appendix 4 of BS 7671 provides current-carrying capacities of cables in certain defined conditions.

Each table specifies the cable type, the ambient temperature, the conductor operating temperature and the reference method of installation (see Table 4A2 of BS 7671). The correct table for any particular cable is found by reference to Table 4A3 of BS 7671.

The tabulated cable current rating I_t is the current that will increase the temperature of the live conductors of the cable from the tabulated ambient (usually 30 °C) to the tabulated maximum conductor operating temperature (e.g. 70 °C for thermoplastic or PVC insulated cables and 90 °C for thermosetting) under the defined conditions, for example see the extract from Table 4D1A.

▼ **Extract from Table 4D1A of BS 7671** Cable current rating table for single-core 70 °C thermoplastic insulated cables, non-armoured, with or without sheath (copper conductors)

CURRENT-CARRYING CAPACITY (amperes):

Ambient temperature: 30 °C
Conductor operating temperature: 70 °C

Conductor cross-sectional area	Reference Method 4 (enclosed in conduit in thermally insulating wall etc.)		Reference Method 3 (enclosed in conduit on a wall or in trunking etc.)		Reference Method 1 (clipped direct)		Reference Method 11 (on a perforated cable tray horizontal or vertical)		Reference Method 12 (free air)		
									Horizontal flat spaced	Vertical flat spaced	Trefoil
	2 cables, single-phase a.c. or d.c.	3 or 4 cables, three-phase a.c.	2 cables, single-phase a.c. or d.c.	3 or 4 cables, three-phase a.c.	2 cables, single-phase a.c. or d.c. flat and touching	3 or 4 cables, three-phase a.c. flat and touching or trefoil	2 cables, single-phase a.c. or d.c. flat and touching	3 or 4 cables, three-phase a.c. flat and touching or trefoil	2 cables, single-phase a.c. or d.c. or 3 cables three-phase a.c.	2 cables, single-phase a.c. or d.c. or 3 cables three-phase a.c.	3 cables trefoil, three-phase a.c.
1	2	3	4	5	6	7	8	9	10	11	12
(mm^2)	(A)	(A)	(A)	(A)	(A)	(A)	(A)	(A)	(A)	(A)	(A)
1	11	10.5	13.5	12	15.5	14	–	–	–	–	–
1.5	14.5	13.5	17.5	15.5	20	18	–	–	–	–	–
2.5	20	18	24	21	27	25	–	–	–	–	–
4	26	24	32	28	37	33	–	–	–	–	–

Example

From the table shown above, for a 4 mm² cable at an ambient temperature of 30 °C enclosed in conduit in an insulating wall $I_t = 26$ A for two cables single-phase a.c. or d.c. and 24 A for three or four cables three-phase a.c.

Rating (correction) factors

If the actual conditions of installation are not as those reference conditions, then rating factors are applied to the tabulated rating I_t to give the cable's current rating I_z in the actual installation conditions, i.e.

$$I_z = I_t \, C_g \, C_a \, C_s \, C_d \, C_i \, C_f \, C_c \qquad\qquad (4.3.1)$$

where:

I_z is the current-carrying capacity of a cable for continuous service under the particular installation conditions concerned

I_t is the tabulated current-carrying capacity of a cable found in Appendix 4 of BS 7671

C_g is the rating factor for grouping, see Tables 4C1 to 4C6 of BS 7671 or Table F3 of the *On-Site Guide*

C_a is the rating factor for ambient temperature, see Tables 4B1 and 4B2 of BS 7671 or Table F1 of the *On-Site Guide*

C_s is the rating factor for thermal resistivity of soil, see Table 4B3 of BS 7671

C_d is the rating factor for depth of buried cable, see Table 4B4 of BS 7671

C_i is the rating factor for conductors surrounded by thermal insulation, see Regulation 523.9 of BS 7671 or Table F2 of the *On-Site Guide*

C_f is the rating factor applied when overload protection is being provided by an overcurrent device with a fusing factor greater than 1.45, e.g. $C_f = 0.725$ for semi-enclosed fuses to BS 3036

C_c is the rating factor for buried circuits: 0.9 for cables buried in the ground requiring overload protection, otherwise is 1.

Table 4B1
Table 4B2

4.3.2 Ambient temperature rating factor C_a

C_a is given by Tables 4B1 and 4B2 of BS 7671. Table 4B1 is used for cables in air, Table 4B2 for cables laid in the ground.

▼ **Table 4B1 of BS 7671**   Rating factors (C_a) for ambient air temperatures other than 30 °C

Ambient temperature[a] (°C)	Insulation			Mineral[a]	
	60°C thermosetting	70°C thermoplastic	90°C thermosetting	Thermoplastic covered or bare and exposed to touch 70°C	Bare and not exposed to touch 105°C
25	1.04	1.03	1.02	1.07	1.04
30	1.00	1.00	1.00	1.00	1.00
35	0.91	0.94	0.96	0.93	0.96
40	0.82	0.87	0.91	0.85	0.92
45	0.71	0.79	0.87	0.78	0.88
50	0.58	0.71	0.82	0.67	0.84
55	0.41	0.61	0.76	0.57	0.80
60		0.50	0.71	0.45	0.75
65		–	0.65	–	0.70
70		–	0.58	–	0.65
75		–	0.50	–	0.60
80		–	0.41	–	0.54
85		–	–	–	0.47
90		–	–	–	0.40
95		–	–	–	0.32

[a] For higher ambient temperatures, consult manufacturer.

▼ **Table 4B2 of BS 7671**   Rating factors (C_a) for ambient ground temperatures other than 20 °C

Ground temperature (°C)	Insulation	
	70 °C thermoplastic	90 °C thermosetting
10	1.10	1.07
15	1.05	1.04
20	1.00	1.00
25	0.95	0.96
30	0.89	0.93
35	0.84	0.89
40	0.77	0.85
45	0.71	0.80
50	0.63	0.76
55	0.55	0.71
60	0.45	0.65
65	–	0.60
70	–	0.53
75	–	0.46
80	–	0.38

Tables 4C1
to 4C6

4.3.3 Group rating factor Cg

For the equation $I_z = I_t\, C_g\, C_a\, C_s\, C_d\, C_i\, C_f\, C_c$, C_g is given by Tables 4C1–4C6 of BS 7671 (see Table 4C1 below); however, where cables are not liable to simultaneous overload see section 4.6.

▼ **Table 4C1 of BS 7671** Rating factors for one circuit or one multicore cable or for a group of circuits, or a group of mulitcore cables, to be used with current-carrying capacities of Tables 4D1A to 4J4A

Item	Arrangement (cables touching)	Number of circuits or multicore cables												To be used with current-carrying capacities, Reference Method
		1	2	3	4	5	6	7	8	9	12	16	20	
1	Bunched in air, on a surface, embedded or enclosed	1.00	0.80	0.70	0.65	0.60	0.57	0.54	0.52	0.50	0.45	0.41	0.38	A to F
2	Single layer on wall or floor	1.00	0.85	0.79	0.75	0.73	0.72	0.72	0.71	0.70	0.70	0.70	0.70	C
3	Single layer multicore on a perforated horizontal or vertical cable tray system	1.00	0.88	0.82	0.77	0.75	0.73	0.73	0.72	0.72	0.72	0.72	0.72	E
4	Single layer multicore on cable ladder system or cleats etc.	1.00	0.87	0.82	0.80	0.80	0.79	0.79	0.78	0.78	0.78	0.78	0.78	

Notes:
1 These factors are applicable to uniform groups of cables, equally loaded.
2 Where horizontal clearances between adjacent cables exceeds twice their overall diameter, no rating factor need be applied.
3 The same factors are applied to:
   – groups of two or three single-core cables;
   – multicore cables.
4 If a group consists of both two- and three-core cables, the total number of cables is taken as the number of circuits, and the corresponding factor is applied to the tables for two loaded conductors for the two-core cables, and to the tables for three loaded conductors for the three-core cables (see also below).
5 If a group consists of n single-core cables it may either be considered as n/2 circuits of two loaded conductors or n/3 circuits of three loaded conductors.
6 The rating factors given have been averaged over the range of conductor sizes and types of installation included in Tables 4D1A to 4J4A and the overall accuracy of tabulated values is within 5%.
7 For some installations and for other methods not provided for in the above table, it may be appropriate to use factors calculated for specific cases, see for example Tables 4C4 and 4C5.
8 Where cables with differing conductor operating temperatures are grouped together, the current rating is to be based upon the lowest operating temperature of any cable in the group.
9 If, due to known operating conditions, a cable is expected to carry not more than 30% of its *grouped* rating, it may be ignored for the purpose of obtaining the rating factor for the rest of the group. For example, a group of N loaded cables would normally require a group rating factor of C_g applied to the tabulated I_t. However, if M cables in the group carry loads which are not greater than $0.3\, C_g I_t$ amperes the other cables can be sized by using the group rating factor corresponding to (N-M) cables.

Grouping factors for three-phase and single-phase circuits in a common enclosure

Table 4C1 provides rating factors for grouping, which apply only to 'uniform groups' that are all single-phase or all three-phase circuits. To estimate grouping factors for groups of single-phase and three-phase circuits, the following method may be used.

To estimate the grouping factor for the single-phase circuits in the group:
   Multiply the number of three-phase circuits by 1.5 (whether four-wire or three-wire doesn't matter),
   and add the result to the number of single-phase circuits,
   to give the equivalent number of single-phase circuits for use with Table 4C1.

To estimate the grouping factor for the three-phase circuits in the group:
   Multiply the number of single-phase circuits by 2/3,
   and add the result to the number of three-phase circuits,
   to give the equivalent number of three-phase circuits for use with Table 4C1.

Example

Two three-phase circuits and two single-phase circuits are installed in a common enclosure for part of their run.

Equivalent number of single-phase circuits = (2 x 3/2) + 2 = 5
Factor C_g from Table 4C1 for the single-phase circuits = 0.60 (and not 0.65 as for 4 similar circuits)
Equivalent number of three-phase circuits = (2 x 2/3) + 2 = 3.3
Factor C_g from Table 4C1 for the three-phase circuits = 0.68 (and not 0.65 as for 4 similar circuits)

When calculating circuit current-carrying capacities (I_z) use the single-phase factor for single-phase circuits and the three-phase factor for three-phase circuits.

Grouping involving lightly loaded circuits

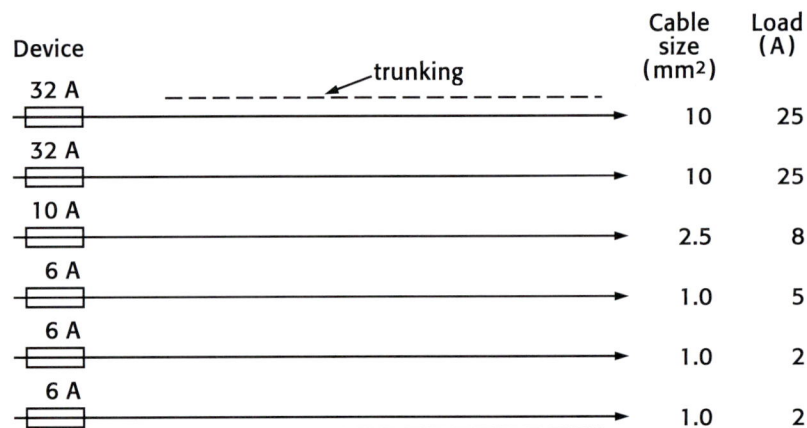

Note 9 to Table 4C1 states:

> If, due to known operating conditions, a cable is expected to carry not more than 30% of its *grouped* rating, it may be ignored for the purpose of obtaining the rating factor for the rest of the group.
>
> For example, a group of N loaded cables would normally require a group rating factor of C_g applied to the tabulated I_t. However, if M cables in the group carry loads which are not greater than 0.3 $C_g I_t$ amperes the other cables can be sized by using the group rating factor corresponding to (N-M) cables.

To see how this is applied, consider the example below:

Example

Six single-phase circuits (2 x 32 A, 1 x 10 A, 3 x 6 A) are wired with single-core thermoplastic insulated cable (Table 4D1A) in a common enclosure for part of their run, see figure below.

**GROUPING INVOLVING
LIGHTLY LOADED CIRCUITS**

Device		Cable size (mm²)	Load (A)
32 A	----- trunking -----	10	25
32 A		10	25
10 A		2.5	8
6 A		1.0	5
6 A		1.0	2
6 A		1.0	2

Factor C_g from Table 4C1 for 6 grouped circuits = 0.57.
If the load of any of the circuits is less than 0.3 $C_g.I_t$ amperes the cable can be discounted in calculating the grouped rating of the other cables.
The rating of the 1.0 mm² cables, from Table 4D1A, is 13.5 A.
Hence 0.3 $C_g.I_t$ for the 6 A circuits is 0.3 x 0.57 x 13.5 = 2.3 A.
If the load of the 6 A circuits is less than 2.3 A, they can be discounted for grouping purposes. In the example illustrated above, the grouping factor for the 4 remaining circuits is 0.65 (and not 0.57 for 6 circuits).

523.9 ### 4.3.4 Thermal insulation factor C_i

C_i is a rating factor for cables run through thermal insulation.

For a single cable surrounded by thermally insulating material over a length of 0.5 m or more, the current-carrying capacity shall be taken, in the absence of more precise information, as 0.5 times the current-carrying capacity for that cable clipped direct to a surface and open (Reference Method C).

Where a cable is to be totally surrounded by thermal insulation for less than 0.5 m, the current-carrying capacity of the cable shall be reduced appropriately depending on the size of cable, length in insulation and thermal properties of the insulation. The derating factors in Table 52.2 are appropriate to conductor sizes up to 10 mm² in thermal insulation having a thermal conductivity greater than 0.04 $Wm^{-1}K^{-1}$.

▼ **Table 52.2 of BS 7671** Cable surrounded by thermal insulation

Length in insulation (mm)	Derating factor
50	0.88
100	0.78
200	0.63
400	0.51

Example

For two 6 mm² thermoplastic insulated cables to Table 4D5 of BS 7671 at an ambient temperature of 25 °C enclosed in a conduit in a thermally insulated wall (Reference Method A):

I_t = 32 A, from Table 4D5
C_g = 0.80 from row 1 of Table 4C1
C_a = 1.03 from Table 4B1
C_i = 1, as no further allowance for insulation is necessary
C_f = 1, as the overcurrent device is not a rewirable fuse
C_c, C_s and C_d = 1, as the cable is not laid in the ground.

Cable rating in installed conditions:
$$I_z = I_t \, C_g \, C_a \, C_i \, C_f \, C_c \, C_s \, C_d$$
$$I_z = 32 \times 0.80 \times 1.03 \times 1 \times 1 \times 1 \times 1 \times 1$$
$$I_z = 26.4 \text{ A}$$

4.3.5 Overcurrent device and buried circuit rating factors C_f, C_c, C_s and C_d

Overcurrent device factor C_f

Most overcurrent devices are assumed to fuse or trip at or below $1.45I_n$, that is I_2 is equal to or less than $1.45I_n$, see Regulation 433.1.1. Where this is the case $C_f = 1$. Where this is not the case and overload protection is being provided, a rating factor must be applied.

$$C_f = 1.45I_n/I_2$$

The most common example is the semi-enclosed (or rewirable) fuse, where I_2 is assumed to have a value of $2I_n$.

Then $C_f = 1.45I_n/2I_n = 0.725$, as given in Regulation 433.1.101.

If no overload protection is provided, whatever the device type, the rating factor C_f is 1. (A semi-enclosed fuse's speed of operation under fault conditions within its fault rating is relatively quick compared with other devices.)

Buried circuit rating factor C_c

The buried cable ratings in Tables 4D4A and 4E4A of BS 7671 are determined at a ground ambient temperature of 20 °C (compared with 30 °C for cables installed in air). Whilst these increased ratings will result in the same conductor operating temperature at full load (70 °C for Table 4D4A), under overload conditions the conductor may exceed the limiting temperature (115 °C for Table 4D4A). The use of the rating factor prevents the cable going over this temperature during an overload, see Regulation 433.1.103.

Soil thermal resistivity rating factor C_s

Soil thermal resistivity affects the rating of a cable laid in the ground, see Table 4B3 of Appendix 4 to BS 7671.

Soil thermal resistivity can be measured with commercially available instruments. If the resistivity is not measured C_s is taken as 0.9.

Depth of laying rating factor C_d

Laying a cable at a depth greater than 700 mm increases the required current rating, see Table 4B4 of BS 7671.

Example

Four 4-core copper armoured cables are laid in touching ducts beneath a road at a depth of 1.25 m. Ground ambient temperature is to be taken as 20 °C and the soil thermal resistivity is measured as 1.0 K.m/W or less. Circuit protection is by BS 88 fuse and an installed rating of 200 A per circuit is required. Determine minimum csa of cable to meet the load requirements. If overload protection is required:

$$I_t \geq \frac{I_z}{C_g C_a C_s C_d C_i C_f C_c}$$

C_g from Table 4C3 is 0.70
C_a from Table 4B2 for a ground ambient of 20 °C is 1

C_s from Table 4B3 for soil thermal resistivity is 1.18

C_d from Table 4B4 is 0.96

C_i is 1

C_f is 1, see Regulation 433.1.100

C_c for a cable laid in the ground is 0.9 when overload protection is required, see Regulation 433.1.102.

$$I_t \geq \frac{200}{0.7 \times 1 \times 1 \times 1 \times 0.9 \times 1.18 \times 0.96}$$

$$I_t \geq 280 \text{ A}$$

From Table 4D4A, column 7 for reference method D, the minimum csa is 240 mm², or 185 mm² if 90 °C thermosetting insulated cables are selected (Table 4E4A), but see 4.3.6 below.

512.1.5 **4.3.6 Conductor operating temperature**

Unless specified by the manufacturer, conductors operating at a temperature exceeding 70 °C are not suitable or safe for use with wiring accessories, low voltage switchgear and controlgear assemblies or other types of equipment, Regulation 512.1.5 refers.

However, 90 °C rated cable can be used provided the conductor operating temperature does not exceed 70 °C, i.e. where the electrical design is based on current ratings given in the equivalent table for 70 °C thermoplastic insulated cables.

432.1 **4.4   Protection against overload and short-circuit**

4.4.1   General

Where the overcurrent device (fuse or circuit-breaker) is providing protection against overload, as is usual, then the cable rating in installed conditions, I_z, must be equal to or greater than the fuse or circuit-breaker rating, I_n.

So, $I_z \geq I_n$, substituting for I_z

$I_t.C_g.C_a.C_s.C_d.C_i.C_f.C_c \geq I_n$

Tabulated rating $I_t \geq I_n/C_g.C_a.C_s.C_d.C_i.C_f.C_c$ (4.4.1)

Example

Consider five three-phase circuits in a common trunking fixed to a wall and unenclosed, using single-core thermoplastic insulated cables to Table 4D1A, in an ambient of 25 °C. Load is 17 A per phase. Type B circuit-breakers are to be used:

A 20 A device is selected
From Table 4B1, $C_a = 1.03$
$C_g = 0.60$ from row 1 column 5 of Table 4C1
$C_i = 1$, as no thermal insulation
C_s, C_d, C_f, $C_c = 1$ as type B devices and the cables are not installed in the ground
$I_t \geq I_n/C_g.C_a.C_s.C_d.C_i.C_f.C_c$
$I_t \geq 20/(0.60 \times 1.03 \times 1 \times 1 \times 1 \times 1 \times 1) = 32.4$ A
so, from Table 4D1A column 5, S = 6 mm²

433.4
523.7

4.4.2 Overcurrent protection of conductors in parallel

Identical cables in parallel are allowed by BS 7671, see Regulation 523.7. They are commonly used for high current LV supplies to main switchboards when single cables of sufficient current-carrying capacity are not available, or to facilitate installation.

The procedure to be applied is:

1 Determine I_b
2 Select I_n (device type and rating), so that $I_n \geq I_b$
3 Using Table 4C1, knowing the reference method and the number of circuits (n), determine the group rating factor C_g
4 Select cables such that $I_t \geq I_n/n.C_g.C_a.C_s.C_d.C_i.C_f.C_c$
5 Refer to appropriate table in Appendix 4 to size the cable conductors.

Example

Consider a 760 A load, to be supplied by two single-core copper 90 °C thermosetting insulated non-armoured cables per phase, all installed on horizontal tray and touching:

1 $I_b = 760$ A
2 Device type and rating: 800 A MCCB, so $I_n = 800$ A
3 If two cables per phase are to be used, for cable from Table 4C5 for reference method F (Note 5), $C_g = 0.91$
   As cables are not to be laid in the ground, ambient is assumed to be 30 °C, $C_a, C_s, C_d, C_i, C_f, C_c = 1$
4 $I_t \geq I_n/n.C_g = 800/(2 \times 0.91) = 440$ A
5 Table 4E1A, column 9 gives 150 mm^2 with a rating of 464 A.

Therefore, 2 x 150 mm^2 cables per phase and neutral (if any) would be required.

433.3
4.5 Protection against fault current only (omission of overload protection)

Note: There are many circumstances where the overcurrent device will be providing fault protection but not overload protection. The most common is when the load is fixed or overload protection is provided by a motor starter. In these circumstances the cable rating must exceed the load but not necessarily the overcurrent device rating.

The general equation where overload protection is not required is:

$$I_t \geq I_b/C_g.C_a.C_s.C_d.C_i.C_f.C_c \tag{4.5.1}$$

(C_f and C_c will be 1, see section 4.3.5 and para 4 of Appendix 4 of BS 7671; see also section 4.6 on non-simultaneous overload.)

Example

As the example of section 4.4.1, except that overload protection is not necessary.

$$I_t \geq I_b/C_g.C_a.C_s.C_d.C_i.C_f.C_c$$

$$I_t \geq 17/(0.60 \times 1.03 \times 1 \times 1 \times 1 \times 1 \times 1) = 27.5\ A$$

so, from Table 4D1A column 5, S = 4 mm²

This equation is appropriate for motor circuits where the motor starter provides overload protection, and for circuits supplying fixed loads. BS 7671 allows its use for any fixed loads, for example water heaters; however, it is usual to provide overload protection unless it is impracticable (Regulation 433.2). (See section 4.7 for motor circuits.)

The omission of overload protection is also allowed where unexpected disconnection would cause danger, see Regulation 433.3.3.

If overload protection is not provided then the adiabatic check of Regulation 434.5.2 is carried out. This is the same check as for protective conductors (Chapter 8).

$$S \geq \frac{\sqrt{I^2t}}{k} \text{ (is a transposition of } t = \frac{k^2S^2}{I^2} \text{ to determine S)}$$

where:

  S   is the cross-sectional area of conductor in mm²
  I   is the effective fault current in amperes, expressed for a.c. as the rms value, due account being taken of the current-limiting effect of the circuit impedances
  t   is the duration in seconds of the fault current
  k   is a factor taking account of the resistivity, temperature coefficient and heat capacity of the conductor material, and the appropriate initial and final temperatures. For the common materials see Table 43.1 of BS 7671.

I is obtained from $I = U_0/Z_s$, t from the device characteristics in Appendix 3 of BS 7671 or the manufacturer and k from Table 43.1 of BS 7671.

The adiabatic equation may not convey any immediate understanding. However, the equation can be rearranged to more clearly demonstrate its objective, as follows:

$$I^2t \leq k^2S^2$$

where:

  I^2t is proportional to the thermal energy let through by the protective device under fault conditions
  k^2S^2 is the thermal capacity of the conductor.

▼ **Figure 4.1** Let-through energy of overcurrent devices

(a) Example of the I^2t characteristic of a fuse

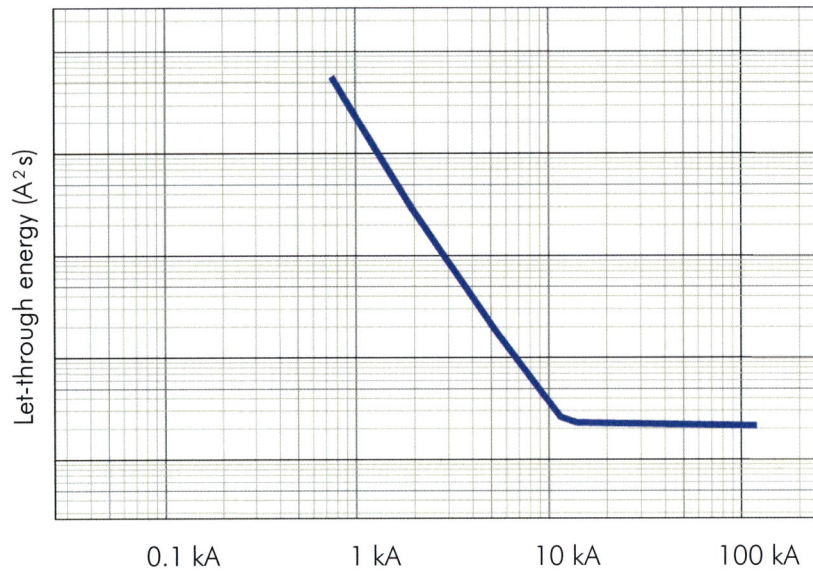

(b) Example of the I^2t characteristic of a circuit-breaker

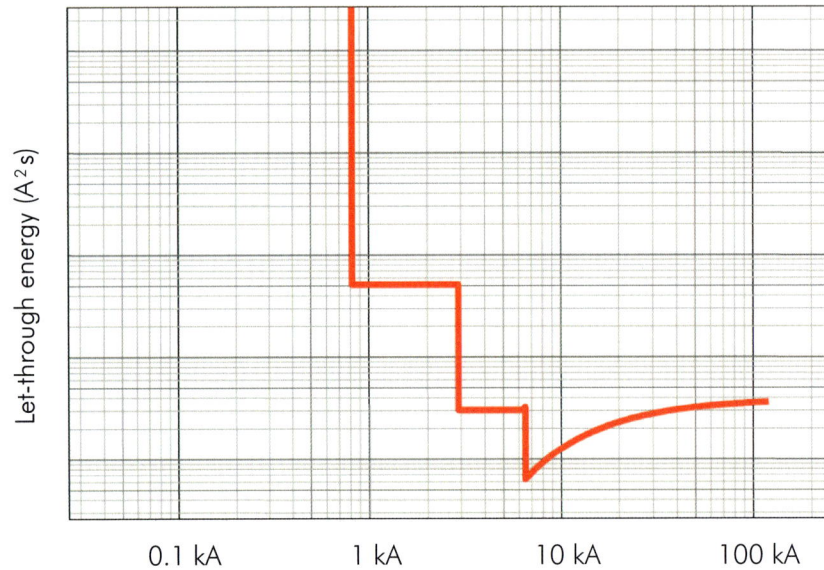

Appx 4 ## 4.6 Corrections for grouping not liable to simultaneous overload

Corrections for grouping are usually made as described above in 4.3.3. However, if the cables are liable to overload but not liable to simultaneous overload, instead of equation 4.4.1 ($I_t \geq I_n/C_g.C_a.C_s.C_d.C_i.C_f.C_c$) the use (of both) of the following two formulae may allow a smaller cable to be used:

$$I_t \geq \frac{I_b}{C_g} \quad \text{and} \quad I_t \geq \sqrt{I_n^2 + 0.48I_b^2 \left(\frac{1 - C_g^2}{C_g^2} \right)}$$

If there are corrections also to be made for ambient temperature, insulation, buried cables and say rewirable fuses, the equations become

$$I_t \geq \frac{I_b}{C_g C_a C_s C_d C_i C_f C_c} \quad \text{and} \quad I_t \geq \frac{1}{C_a C_s C_d C_i} \sqrt{\left(\frac{I_n}{C_f C_c}\right)^2 + 0.48 I_b^2 \left(\frac{1-C_g^2}{C_g^2}\right)}$$

i.e. I_t must be greater than or equal to the larger of the two.

Most cable groupings are unlikely to be liable both to individual and to simultaneous overload.

Example

Consider five three-phase circuits liable to overload but not simultaneously in a common trunking fixed to a wall and unenclosed, using single-core thermoplastic insulated cables to Table 4D1A, in an ambient of 25 °C. Load is 17 A per phase. Type B circuit-breakers are to be used:

A 20 A device is selected
From Table 4B1, $C_a = 1.03$
$C_g = 0.60$ from row 1 column 5 of Table 4C1
$C_i = 1$, as no thermal insulation
$C_c = 1$, as type B circuit-breakers
C_s, C_d and $C_f = 1$, as cables are not installed in the ground

$$I_t \geq \frac{I_b}{C_g C_a C_s C_d C_i C_f C_c} \quad \text{and} \quad I_t \geq \frac{1}{C_a C_s C_d C_i} \sqrt{\left(\frac{I_n}{C_f C_c}\right)^2 + 0.48 I_b^2 \left(\frac{1-C_g^2}{C_g^2}\right)}$$

$$I_t \geq \frac{17}{0.6 \times 1.03 \times 1 \times 1 \times 1 \times 1 \times 1} = 27.5 \text{ A} \quad \text{and}$$

$$I_t \geq \frac{1}{1.03 \times 1 \times 1 \times 1} \sqrt{\left(\frac{20}{1 \times 1}\right)^2 + 0.48 \times 17^2 \left(\frac{1-0.6^2}{0.6^2}\right)} = 24.7 \text{ A}$$

I_t must be equal to or greater than 27.5 A and from Table 4D1A 4 mm^2 cable can be used.

4.7 Motors

▼ **Figure 4.2**
Motor with star/delta starting

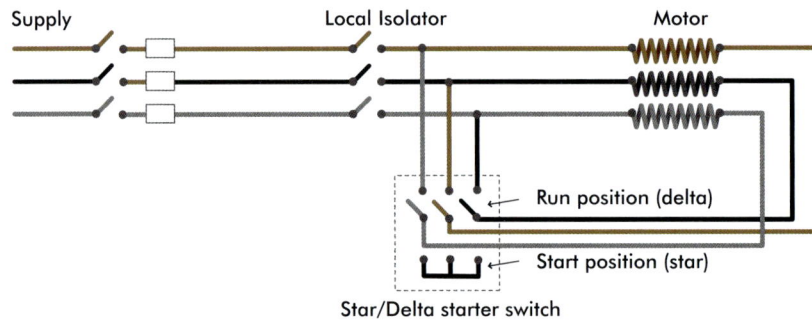

The starting current for direct-on-line starting can be up to 8 times full-load current, with a starting torque of 1.5 to 2 times full-load torque.

Star/delta starting can reduce starting currents. The machine is connected in star whilst running up to speed, then switches to delta connection. The line current is 1/3 that in delta, but starting torque is reduced to 1/3.

The cable supplying the motor starter/local isolator is sized to take the motor load (continuous rating).

$$I_b = \frac{\text{motor kVA} \times 1000}{\sqrt{3} \times U} \quad \text{or} \quad I_b = \frac{\text{motor kW} \times 1000}{\sqrt{3} \times U \times \cos \varnothing}$$

where:

    kVA is the motor rating
    U   is the voltage between lines
    kW is the motor rating in kW
    cos Ø   is the motor power factor.

Because of high starting currents the overcurrent devices will be motor rated, that is, of such rating that they will withstand the starting surges.

Example 1

Consider a 16 kW motor at 400 V with a full-load power factor of 0.8.

$$I_b = \frac{\text{motor kW} \times 1000}{\sqrt{3} \times U \times \cos \varnothing} = \frac{16 \times 1000}{\sqrt{3} \times 400 \times 0.8} = 28.9 \text{ A}$$

Full-load current I_b is 29 A and a 32M50 motor fuse to BS 88 (BS EN 60269-2) is selected to prevent fusing on starting.

Example 2

Assume a four-core armoured thermoplastic insulated cable 80 m in length is to be used and it will be clipped to a perforated cable tray. Table 4D4A of BS 7671 indicates a 4 mm² four-core cable would be suitable. It is proposed to use the cable armour as the circuit protective conductor.

A 32M50 fuse has a continuous current rating of 32 A, but behaves otherwise as a 50 A fuse.

Clearly, the 32M50 will not provide overload protection; this is provided by the motor starter.

It needs to be confirmed that the fuse will protect both the conductors and the armour of the cable in the event of a fault.

Let us assume the motor is to be supplied from distribution board C of Figure 6.4, in section 6.3.2.

	Impedance	
	r (Ω)	x (Ω)
Earth loop impedance at C from 6.3.3 para 3	0.0815	0.0597
80 m of 4 mm² 4-core Cu SWA		
(a) line impedance at 20 °C col 3 of Table F.1 80 x 4.61/1000	0.3688	–
(b) armour impedance at 20 °C col 6 of Table F.7A		
80 x 4.60/1000	0.3680	
(c) correction of (a) to 70 °C (x 0.20) (col 3 Table F.17)	0.0738	
(d) correction of (b) to 60 °C (x 0.18) (col 3 Table F.17)	0.0662	
Total loop impedance at motor starter	0.9583	0.0597

Note: For the extensive calculations of a large installation it is simpler to list the impedance at 20 °C and add a correction to the operating temperature, see section 6.3.2.

$$Z_{ef} = \sqrt{0.9583^2 + 0.0597^2} = 0.96 \quad \text{and} \quad I_{ef} = \frac{230}{0.96} = 240 \text{ A}$$

From Figure 3A3 (a) of Appendix 3 of BS 7671 (see Chapter 8), the disconnection time t for a 50 A fuse with a fault current of 240 A is 3 s.

Using the formula $S \geq \dfrac{\sqrt{I^2 t}}{k}$

where k is given by Table 43.1 of BS 7671:

For the line conductors $S \geq \dfrac{\sqrt{240^2 \times 3}}{143} = 2.91 \text{ mm}^2$

The cable core size is 4 mm², so this is satisfactory.

For the armour $S \geq \dfrac{\sqrt{240^2 \times 3}}{52} = 7.99 \text{ mm}^2$.

From column 4 of Table F.7B, the area of the armouring is 35 mm², so this is satisfactory.

Example 3

When a three-phase motor is controlled by a star-delta starter the ends of the three motor windings are extended from the motor terminals to the starter. The connections of the windings in star for start, and delta for run are made at the starter, see Figure 4.2.

552.1 The steady running current (I_w) in each of the six conductors between a star-delta starter and its motor is 1/√3 (58%) of that through the conductors supplying the starter.

I_L (run position, delta connected) $= 2.\cos 30 \times I_w$, where I_w is the winding current.

Hence $I_w = \dfrac{I_L}{2.\cos 30} = \dfrac{I_L}{\sqrt{3}}$

The 6 cables (2 circuits) connecting the motor to the starter are usually run in common conduit or trunking. Table 4C1 (section 4.3.3) gives a group rating factor C_g of 0.8 for such an arrangement.

A procedure for the selection of motor cables is illustrated by the following example.

Assume a full-load motor current I_b of 37 A.

Supply cable, to be three-core thermoplastic (PVC) SWA on a perforated tray, ambient temperature 30 °C (C_g, C_a, C_s, C_d, C_i, C_f, $C_c = 1$).

Motor cables, to be single-core 70 °C thermoplastic (PVC) in conduit, ambient temperature 40 °C ($C_a = 0.87$, Table 4B1), double circuit enclosed ($C_g = 0.8$, Table 4C1).

For the supply cable, Regulation 433.1.1 ($I_n \geq I_b$) is met by a starter for which $I_n = 40$ A.

$I_t \geq I_n/C_g.C_a.C_s.C_d.C_i.C_f.C_c = 40/(1 \times 1 \times 1 \times 1 \times 1 \times 1 \times 1) = 40$ A.

A three-core 6 mm² cable has a tabulated rating, $I_t = 45$ A (column 5 Table 4D4A).

If the delayed overload trips operate at, say, 50 A (I_2), condition (iii) of Regulation 433.1.1,

$I_2 \leq 1.45\,I_z$ where $I_z = I_t.C_g.C_a.C_s.C_d.C_i.C_f.C_c$ (see section 4.2.3 for I_2),

is met because $I_2 = 50$ A, and $1.45 \times 45 \times 1 \times 1 \times 1 \times 1 \times 1 \times 1 \times 1 = 65$ A.

For the motor cables, the procedure requires that

$I_t \geq I_n/\sqrt{3}.C_g.C_a.C_s.C_d.C_i.C_f.C_c = 40/(\sqrt{3} \times 0.87 \times 0.8 \times 1 \times 1 \times 1 \times 1 \times 1) = 33$ A.

6 mm² single-core cables in conduit, which have a single circuit tabulated rating, I_t, of 36 A are selected (see column 5 of Table 4D1A).

If overload protection is satisfactory for the supply cables, the motor cables will also be protected. This can be demonstrated as follows:

$I_2 = 50$ (which is a line current)

For the motor cables, which are a double circuit in conduit,

$I_z = I_t.C_a.C_g.C_i = 36 \times 0.87 \times 0.8 \times 1 = 25$ A

$1.45\,I_z = 1.45 \times 25 = 36.25$ A.

The line equivalent of this current in the delta connected motor cables is $\sqrt{3} \times 36.25 = 63$ A.

Consequently, the 6 mm² motor cables are protected by the overload trips ($I_2 = 50$ A) in the line conductors.

Voltage drop

5

- ■ Consumers' installations
- ■ Distribution system voltage drop
- ■ Basic voltage drop calculation
- ■ Correction for inductance
- ■ Correction for load power factor
- ■ Correction for conductor operating temperature
- ■ Correction for both conductor operating temperature and load power factor

525, Appx 4 sect 6

5.1 Voltage drop in consumers' installations

BS 7671 requires that under normal service conditions the voltage at the terminals of any fixed current-using equipment shall be greater than the lower limit corresponding to the British Standard relevant to the equipment. There is no specific voltage drop requirement as such. The voltage drop should simply not exceed that for the proper working of the equipment.

The requirements are deemed to be satisfied for a supply given in accordance with the Electricity Safety, Quality and Continuity Regulations 2002 (that is, with a supply voltage within the range 230 V plus 10% to 230 V minus 6%) if the voltage drop between the origin of the installation (usually the supply terminals) and the load point is not greater than 3% of the nominal voltage for lighting circuits and 5% for other circuits.

A greater voltage drop may be accepted for a motor during starting periods and for other equipment with high inrush currents provided that it is verified that the voltage variations are within the limits specified in the relevant British Standards for the equipment or, in the absence of a British Standard, in accordance with the manufacturer's recommendations.

Appx 4 sect 6

5.2 Distribution system voltage drop

Large installations cannot be designed to a total voltage drop of 5%. If the voltage at the origin is within the designer's control, as when the supply is from an 11 000 V/433 V transformer, use can be made of the Electricity Safety, Quality and Continuity Regulations allowed voltage range of +10% to −6%. BS 7671 assumes this in setting the deemed to comply 3% or 5% for a supply from a public low voltage system.

Assuming a nominal 230 V line to earth at the furthest consumer, the open circuit or highest voltage could be 230 V plus 10%, that is 230 + 23 = 253 V and the lowest 230 V minus 6%, that is 230 − 13.8 = 216.2 V, giving a total voltage drop of 36.8 V, or 14.5% of 253 V. These would be voltage drops to the origin of the final circuits, allowing the use of standard final circuits with a further 3% or 5% voltage drop.

A conservative limit of 220 V at the point of use would allow 23 V in the distribution (9% of 253 V as in Figure 5.1), and 9.2 V in the final circuit (4% of 230 V) has been adopted in the figure below:

▼ **Figure 5.1** Distribution system voltage drop

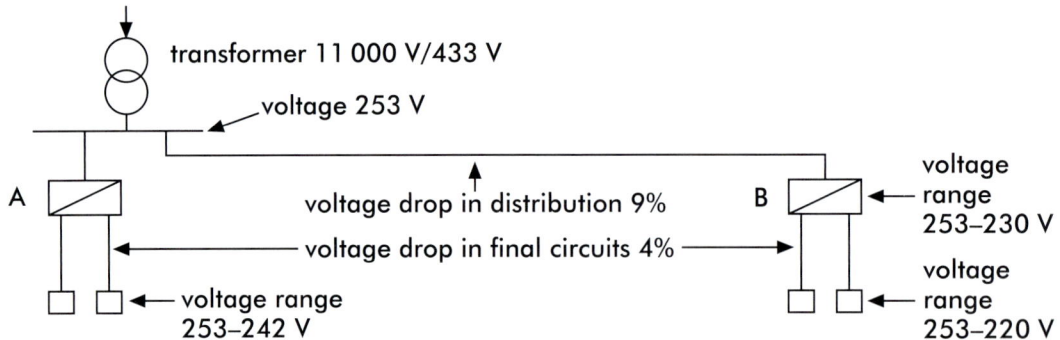

Note: If final circuits were designed to 5%, the voltage range would be 253–218.5 V, and if to 3% then 253–223 V.

Appx 4 sect 6 ## 5.3 Basic voltage drop calculation

5.3.1 Single-phase

To calculate the voltage drop in volts the tabulated value of voltage drop (mV/A/m) from Appendix 4 of BS 7671 has to be multiplied by the design current of the circuit (I_b), the length of run in metres (L), and divided by 1000 (to convert to volts).

$$\text{Voltage drop} = \frac{L.I_b.(\text{mV/A/m})}{1000}$$

where:

(mV/A/m)   is the tabulated value of voltage drop in mV per amp per metre from Appendix 4 of BS 7671

L   is the the length of run in metres (m)

I_b is the design current of the circuit (A).

The requirements of BS 7671 are deemed to be satisfied for a 230 V supply, if the voltage drop between the origin of the installation and a socket-outlet or fixed current-using equipment does not exceed at full load 5% of 230 V, that is 11.5 V, or 3% for lighting, that is 6.9 V.

Example

Consider a cooker circuit with I_b of 30 A, installed in conduit in an insulated wall; from Table 4D5 method A column 7, a 6 mm² cable is selected. From column 8 this has a voltage drop (per ampere per metre) of 7.3 mV/A/m. If the length of cable is 10 m then the voltage drop is

$$\frac{L.I_b.(\text{mV/A/m})}{1000} = \frac{10 \times 30 \times (7.3)}{1000} = 2.19 \text{ V}$$

This is acceptable as it is less than 11.5 V.

5.3.2 Three-phase voltage drop

The tables of Appendix 4 of BS 7671 give voltage drop data for two-core cables (for example, columns 2 and 3 of Table 4D4B), that is d.c. and single-phase a.c., and for three- and four-core cables (column 4 of Table 4D4B), that is three-phase a.c.

The same voltage drop is given for three- and four-core cables, as the presumption is that it is a balanced load with no triple harmonics so that the current in the neutral of the four-core cable is zero. For a balanced three-phase load no neutral is required.

The voltage drop for two-core cables (single-phase) is the voltage drop in the line to earth voltage, say 230 V.

The voltage drop for three- and four-core cables (three-phase) is the voltage drop in the line-to-line voltage, say 400 V.

By inspection of the extract from Table 4D4B below, it can be seen that the three- and four-core cable voltage drop per amp per metre (mV/A/m) is $\frac{\sqrt{3}}{2}$ times the two-core (mV/A/m). The $\sqrt{3}$ converts to three-phase; the division by two is necessary as there is assumed to be no voltage drop in the neutral.

Example

Consider a three-phase 10 kVA motor circuit wired with three-core armoured thermoplastic insulated cable with a circuit length 10 m.

Now load in kVA $= 3.U_0.I_b /1000$ or $\sqrt{3}.U.I_b /1000$, (as $3.U_0 = \sqrt{3}.U$)

hence $I_b = $ (load in kVA) x $1000/3.U_0 = 10$ x $1000/3/230 = 14.49$ A

where:

    U   voltage between lines (V)
    U_0 nominal line voltage to earth (V)
    I_b design current of circuit (A).

From Table 4D4A method E column 5, a 1.5 mm² cable is selected.

From Table 4D4B column 4, as we have a three-phase supply there is a voltage drop (per ampere per metre) of 25 mV/A/m so the line voltage drop is

$$\frac{L.I_b.(mV/A/m)}{1000} = \frac{10 \times 14.49 \times (25)}{1000} = 3.62 \text{ V}$$

The percentage voltage drop is (3.62/400) x 100% = 0.9%

▼ Extract from Table 4D4B of BS 7671

VOLTAGE DROP (per ampere per metre):   Conductor operating temperature: 70°C

Conductor cross-sectional area	Two-core cable, d.c.	Two-core cable, single-phase a.c.			Three- or four-core cable, three-phase a.c.		
1	2	3			4		
(mm²)	(mV/A/m)	(mV/A/m)			(mV/A/m)		
1.5	29	29			25		
2.5	18	18			15		
4	11	11			9.5		
6	7.3	7.3			6.4		
10	4.4	4.4			3.8		
16	2.8	2.8			2.4		
		r	x	z	r	x	z
25	1.75	1.75	0.170	1.75	1.50	0.145	1.50
35	1.25	1.25	0.165	1.25	1.10	0.145	1.10
50	0.93	0.93	0.165	0.94	0.80	0.140	0.81
70	0.63	0.63	0.160	0.65	0.55	0.140	0.57
95	0.46	0.47	0.155	0.50	0.41	0.135	0.43

5.3.3 Summing voltage drop

Adding percentage voltage drops

There is an advantage in expressing voltage drops in percentage terms, in that they can then be added together directly, e.g. 2% in three-phase distribution plus 2% in either single-phase or three-phase final circuits is 4% overall in both cases.

Summing voltage drops

To add three-phase voltage drops expressed in volts to single-phase voltage drops in volts, the three-phase voltage drop must be divided by $\sqrt{3}$.

Example

Consider a 400 V distribution cable with a line-to-line voltage drop of 5% and a single-phase circuit supplied by the distribution cable with a line to neutral voltage drop of 4%.

What is the overall voltage drop to the end of the single-phase circuit?

Adding percentage voltage drops: 5% + 4% = 9%, or 230 x 9/100 = 20.7 V

Summing voltage drops: 5% of 400 V is 20 V line voltage drop, divide by $\sqrt{3}$ to give phase voltage drop of $20/\sqrt{3}$ = 11.55 V. 4% of 230 V is 9.2 V. Total voltage drop is 11.55 + 9.2 = 20.75 V. (They are not exactly the same as $400/\sqrt{3}$ is not exactly 230.)

Appx 4 sect 6 ## 5.4 Correction for inductance

For cables having conductors of 16 mm² or less cross-sectional area their inductances are not significant and $(mV/A/m)_r$ values only are tabulated. For cables having conductors greater than 16 mm² cross-sectional area, the impedance values are given as $(mV/A/m)_z$, together with the resistive component $(mV/A/m)_r$ and the reactive component $(mV/A/m)_x$, see the extract from Table 4D4B given earlier and Table 4H4B below.

Calculations for Electricians and Designers
© The Institution of Engineering and Technology

All cables have resistance and inductance and the inductance is quite similar for all cable sizes. However, for cables 16 mm² and smaller, values of inductance are not provided as it is negligible compared with the resistance.

If the power factor of the load is not known, the z value of voltage drop should be used. When corrections are to be made for load power factor, r and x values are used, see section 5.5.

Where the load power factor is not known, voltage drop is determined using the formula:

$$\text{Voltage drop} = \frac{L.I_b.(mV/A/m)_z}{1000}$$

▼ Table 4H4B of BS 7671

VOLTAGE DROP (per ampere per metre): Conductor operating temperature: 70°C

Conductor cross-sectional area	Two-core cable, d.c.	Two-core cable, single-phase a.c.			Three- or four-core cable, three-phase a.c.		
1	2	3			4		
(mm²)	(mV/A/m)	(mV/A/m)			(mV/A/m)		
16	4.5	4.5			3.9		
		r	x	z	r	x	z
25	2.9	2.9	0.175	2.9	2.5	0.150	2.5
35	2.1	2.1	0.170	2.1	1.80	0.150	1.80
50	1.55	1.55	0.170	1.55	1.35	0.145	1.35
70	1.05	1.05	0.165	1.05	0.90	0.140	0.92
95	0.77	0.77	0.160	0.79	0.67	0.140	0.68
120	–	–	–	–	0.53	0.135	0.55
150	–	–	–	–	0.42	0.135	0.44
185	–	–	–	–	0.34	0.135	0.37
240	–	–	–	–	0.26	0.130	0.30
300	–	–	–	–	0.21	0.130	0.25

Example

Consider a three-phase distribution circuit with an estimated load I_b of 300 A and a cable route length of 100 m. It has been decided that the voltage drop should not exceed 9%, see section 5.2. From Table 4H4A a 240 mm² aluminium cable has been selected and from Table 4H4B the (mV/A/m) from column 4 is r = 0.26, x = 0.130, z = 0.30 mV/A/m.

Note: $z = \sqrt{r^2 + x^2}$

If the power factor of the load is not known the z value of (mV/A/m) is used, that is 0.30 in this example.

9% of the line-to-line voltage of 400 V is 36 V, hence

the voltage drop at design load would be $\dfrac{L.I_b.(mV/A/m)_z}{1000} \leq 36$

$$\frac{L.I_b.(mV/A/m)_z}{1000} = \frac{100 \times 300 \times (0.30)}{1000} = 9\ V$$

This is acceptable.

Where the load power factor is known, voltage drop is determined using the formula in section 5.5.

Appx 4 sect 6.2 ## 5.5 Correction for load power factor

The use of the tabulated mV/A/m values, and for cable sizes over 16 mm² the tabulated $(mV/A/m)_z$ values, to calculate the voltage drop is strictly correct only when the phase angle of the cable equals that of the load. When the phase angle of the cable does not equal that of the load, the direct use of the tabulated mV/A/m or $(mV/A/m)_z$ values leads to a calculated value of voltage drop higher than the actual value.

Where a more accurate assessment of voltage drop is required the following methods may be used.

For cables having conductors of cross-sectional area of 16 mm² or less, the design value of mV/A/m is obtained approximately by multiplying the tabulated value by the power factor of the load, cos Ø, i.e.

$$\text{Voltage drop} = \frac{L.I_b.(mV/A/m) \cos \emptyset}{1000}$$

For cables having conductors of cross-sectional area greater than 16 mm² the design value of mV/A/m is given approximately by:

tabulated $(mV/A/m)_r \cos \emptyset$ + tabulated $(mV/A/m)_x \sin \emptyset$, and

$$\text{Voltage drop} = L.I_b.\frac{(mV/A/m)_r \cos \emptyset + (mV/A/m)_x \sin \emptyset}{1000}$$

Example

Consider a three-phase distribution circuit with an estimated load I_b of 300 A with a power factor cos Ø of 0.8 and a cable route length of 100 m. It has been decided that the voltage drop should not exceed 9%, see section 5.2. From Table 4H4A a 240 mm² aluminium cable has been selected and from Table 4H4B the (mV/A/m) from column 4 is r = 0.26, x = 0.13, z = 0.30 mV/A/m.

$$\text{Voltage drop} = L.I_b.\frac{(mV/A/m)_r \cos \emptyset + (mV/A/m)_x \sin \emptyset}{1000}$$

$$\text{Voltage drop} = 100 \times 30 \times \frac{(0.26 \times 0.8) + (0.13 \times 0.6)}{1000} = 8.58 \text{ V}$$

For single-core cables in flat formation the tabulated values apply to the outer cables and may underestimate for the voltage drop between an outer cable and the centre cable for cross-sectional areas above 240 mm², and power factors lower than 0.8.

Appx 4 sect 6.1 ## 5.6 Correction for conductor operating temperature

Where the design current of a circuit is significantly less than the effective current-carrying capacity of the chosen cable, the actual voltage drop would be less than the calculated value because the conductor temperature (and hence its resistance) will be less than that on which the tabulated (mV/A/m) is based.

C_t is a correction factor that can be applied if the load current is significantly less than I_z, the current-carrying capacity of the cable in the particular installation conditions; that is, if $I_b < I_t.C_g.C_a.C_i$. If C_t is taken as 1, any error will be on the safe side. This factor compensates for the temperature of the cable at the reduced current being less than the temperature at the maximum current. Because resistance is dependent upon temperature, there is a reduction in resistance if the cable is not fully loaded and the voltage drop is correspondingly reduced. If required, further information can be found in the *Commentary*.

For cables having conductors of cross-sectional area 16 mm^2 or less, the design value of (mV/A/m) is obtained by multiplying the tabulated value by a factor C_t, given by

$$C_t = \frac{230 + t_p - \left(C_g{}^2\, C_a{}^2\, C_s{}^2\, C_d{}^2 - \dfrac{I_b{}^2}{I_t{}^2}\right)(t_p - 30)}{230 + t_p}$$

where t_p is the maximum permitted normal conductor operating temperature (°C).

This factor should only be applied to the resistive element of the (mV/A/m).

This equation strictly only applies where the overcurrent protective device is other than a BS 3036 fuse and where the actual ambient temperature is equal to or greater than 30 °C.

Note: For convenience, the above formula is based on the approximate resistance-temperature coefficient of 0.004 per °C at 20 °C for both copper and aluminium conductors.

For conductors of 16 mm^2 csa or less:

$$\text{Voltage drop} = \frac{L.I_b.(mV/A/m).C_t}{1000}$$

For conductors of csa greater than 16 mm^2:

$$\text{Voltage drop} = \frac{L.I_b.\sqrt{((mV/A/m)_r\,.C_t)^2 + (mV/A/m)_x{}^2}}{1000}$$

Example

Consider a 20 A radial socket-outlet circuit to be wired in 2.5/1.5 mm² flat twin with cpc cable per Table 4D5, with the cable clipped direct, installation method C. Assume $C_a = C_g = 1$.

From Table 4D5,

$t_p = 70$, $I_b = 20$, $I_t = 27$, (mV/A/m) = 18

Then $C_t = \dfrac{230 + 70 - \left(1^2\ 1^2\ 1^2\ 1^2 - \dfrac{20^2}{27^2}\right)(70 - 30)}{230 + 70} = 0.94$

For a voltage drop of 5% of 230 V, that is 11.5 V, the maximum circuit length L is derived from

Voltage drop $= \dfrac{L.I_b.(mV/A/m).C_t}{1000}$; hence $\dfrac{L \times 20 \times (18) \times 0.94}{1000}$, thus $L = 34$ m

If the installation method is A then $I_t = 20$ A, $C_t = 1$ and maximum length is 32 m.

For cables having conductors of cross-sectional area greater than 16 mm², only the resistive component of the voltage drop is affected by the temperature and the factor C_t is therefore applied only to the tabulated value of $(mV/A/m)_r$; the design value of $(mV/A/m)_z$ is given by the vector sum of C_t $(mV/A/m)_r$ and $(mV/A/m)_x$.

For very large conductor sizes where the resistive component of voltage drop is much less than the corresponding reactive part (i.e. when $x/r \geq 3$) this correction factor need not be considered.

Appx 4 sect 6.3

5.7 Correction for both conductor operating temperature and load power factor

From sections 5.5 and 5.6 above, to correct the tabulated (mV/A/m) values for both conductor operating temperature and load power factor, the design values of (mV/A/m) are given by:

for cables having conductors of 16 mm² or less cross-sectional area

tabulated (mV/A/m) cos Ø.C_t

for cables having conductors of cross-sectional area greater than 16 mm²

tabulated $(mV/A/m)_r$ cos Ø.C_t + tabulated $(mV/A/m)_x$ sin Ø, and

Voltage drop $= \dfrac{(mV/A/m)_r \cos Ø.C_t + (mV/A/m)_x \sin Ø}{1000} L.I_b$

Calculation of fault current

6

- ■ **Determination of prospective fault current**
- ■ **Determined by enquiry**
- ■ **Determined by calculation**

6.1 Determination of prospective fault current

434.1 Regulation 434.1 requires the prospective fault current (i.e. under both short-circuit and earth fault conditions) to be determined at every relevant point of the installation. This means that at every point where switchgear is installed, the maximum fault current must be determined to ensure that the switchgear is adequately rated to interrupt any fault currents which may occur on its load side, see Table 6.1.

▼ **Table 6.1** Rated short-circuit capacities

Device type		Device designation	Rated short-circuit capacity (kA)	
Semi-enclosed fuse to BS 3036 with category of duty		S1A	1	
		S2A	2	
		S4A	4	
Cartridge fuse to BS 1361	type I		16.5	
	type II		33.0	
Cartridge fuse to BS 88-2 Fuse system E (bolted) Fuse system G (clip-in) Fuse system gU			80 kA a.c., 40 kA d.c. 50 kA size E1, 80 kA sizes F1, F2, F3, 50 kA	
Cartridge fuse to BS 88-3	type 1		16	
	type 2		31.5	
Cartridge fuse to BS 88-6			16.5 at 240 V 80 at 415 V	
Circuit breakers to BS 3871 (replaced by BS EN 60898)		M1	1	
		M1.5	1.5	
		M3	3	
		M4.5	4.5	
		M6	6	
		M9	9	
Circuit-breakers to BS EN 60898 and RCBOs to BS EN 61009			I_{cn}	I_{cs}
			1.5	(1.5)
			3.0	(3.0)
			4.5	(4.5)
			6	(6.0)
			10	(7.5)
			15	(7.5)
			20	(10.0)
			25	(12.5)

Notes to Table 6.1:

Two short-circuit ratings are defined in BS EN 60898 and BS EN 61009:

> I_{cn} is the rated short-circuit capacity (marked on the device)
> I_{cs} is the service short-circuit capacity.

The difference between the two is the condition of the circuit-breaker after manufacturer's testing.

I_{cn} is the maximum fault current the breaker can interrupt safely, although the breaker may no longer be usable.

I_{cs} is the maximum fault current the breaker can interrupt safely without loss of performance.

The I_{cn} value (in amperes) is normally marked on the device in a rectangle, e.g. $\boxed{6000}$ and for the majority of applications the prospective fault current at the terminals of the circuit-breaker should not exceed this value.

For domestic installations the prospective fault current is unlikely to exceed 6 kA, up to which value the I_{cn} and I_{cs} values are the same.

The short-circuit capacity of devices to BS EN 60947-2 is as specified by the manufacturer.

Also, to ensure that cables are properly protected, it is necessary to calculate the lowest fault current on each cable run, e.g. at the extremity. This is for two purposes:

i   to confirm that there is sufficient fault current to cause operation of the overcurrent device in the event of a fault in the required disconnection time to provide protection against electric shock

ii   to confirm that during faults the energy let through by the device does not damage the cable.

Protective devices let most energy through at low fault currents (relative to the overcurrent device rating), and as a consequence the highest conductor temperature will occur for the minimum fault current because the disconnection time is greatest.

Regulation 434.1 requires that:

> **The prospective fault current shall be determined at every relevant point of the installation. This shall be done by calculation, measurement or enquiry.**

Note: The wording of the Electrical Installation Certificate recognizes this choice.

6.2 Determined by enquiry

6.2.1 General

BS 7671 makes specific reference to enquiry as a method of determining the fault current.

Regulation 28 of the Electricity Safety, Quality and Continuity Regulations 2002 states:

Information to be provided on request

28. A distributor shall provide, in respect of any existing or proposed consumer's installation which is connected or is to be connected to his network, to any person who can show a reasonable cause for requiring the information, a written statement of –

(a) the maximum prospective short-circuit current at the supply terminals;

(b) for low voltage connections, the maximum earth loop impedance of the earth fault path outside the installation;

(c) the type and rating of the distributor's protective device or devices nearest to the supply terminals;

(d) the type of earthing system applicable to the connection; and

(e) the information specified in regulation 27(1),

which apply, or will apply, to that installation.

6.2.2 Enquiry – maximum prospective short-circuit current, single-phase supplies up to 100 A

Electricity distributors will generally declare a maximum prospective short-circuit current at the distributor's cut-out of 16 kA (0.55 p.f.), see Table F.18A. The fault level will only be this high if the installation is close to the distribution transformer. However, over the lifetime of an installation, changes may be made to the distribution network and consequently designers must install equipment suitable for the highest fault levels which might occur. Attenuation or reduction of these fault levels may be estimated for that part of the service line on the customer's premises on the assumption that, whilst the distribution network on the public highway might change, the service line on the customer's premises will remain unchanged (or at least not be changed without the occupant's knowledge). Therefore, attenuation should only be allowed for cable on the customer's premises. Within these constraints, attenuation as per Table F.18A can be made. However, as consumer units and fusegear manufactured to British Standards can be obtained with a conditional rating of 16 kA, the selection of such switchgear eliminates the need for such allowances or estimates to be made.

There are some inner city locations where the maximum prospective short-circuit current on the distributing main exceeds 16 kA. The distributor, in making its declaration under the Electricity Safety, Quality and Continuity Regulations, should advise of any such situations.

6.2.3 Enquiry – maximum prospective fault current, three-phase supplies

Electricity distributors will provide estimated maximum prospective fault currents (p.f.c.) at the cut-out of three-phase supplies, based on either:

a a declared level of 25 kA (0.23 p.f.) at the point of connection of the service line to the busbars in the distribution substation, see Table F.18B, or

b a declared level of 18 kA (0.5 p.f.) at the point of connection of the service line to the low voltage distribution main, see Table F.18C.

Knowledge of the power factor (p.f.) as well as the p.f.c. enables a more accurate calculation of p.f.c. downstream of the supply to be carried out. The p.f.c. can be resolved into resistive and reactive components.

Information on the distribution network is necessary to use the attenuation tables, and Tables F.18A to C (Appendix F) are provided for this. Attenuation should only be allowed for the length of service line on the customer's premises. Figure 6.1 is intended to demonstrate this. It is necessary to agree likely fault levels with the electricity distributor.

▼ **Figure 6.1** Distribution and service cables

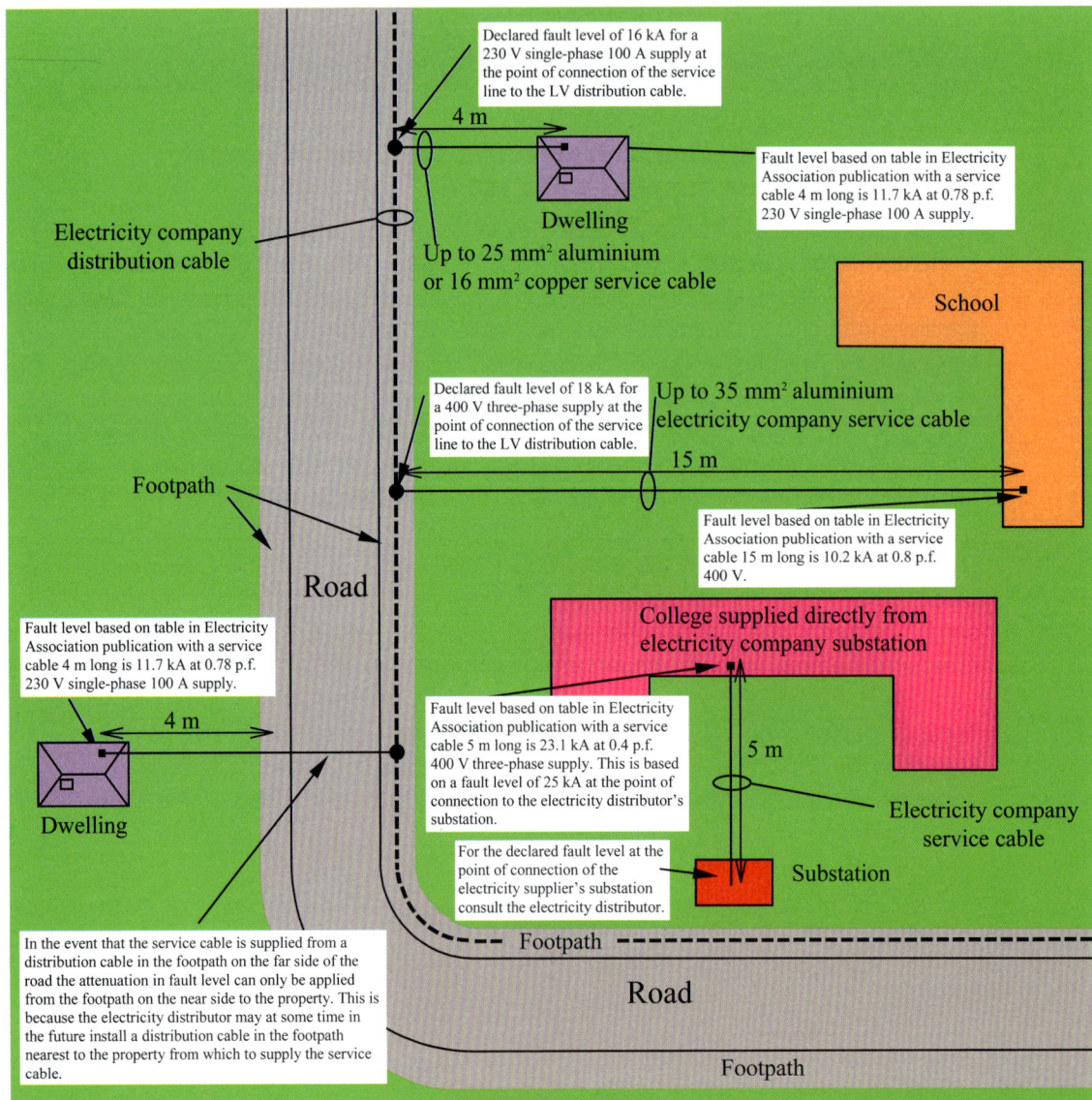

Declared fault level of 16 kA for a 230 V single-phase 100 A supply at the point of connection of the service line to the LV distribution cable.

4 m

Fault level based on table in Electricity Association publication with a service cable 4 m long is 11.7 kA at 0.78 p.f. 230 V single-phase 100 A supply.

Dwelling

Electricity company distribution cable

Up to 25 mm² aluminium or 16 mm² copper service cable

School

Declared fault level of 18 kA for a 400 V three-phase supply at the point of connection of the service line to the LV distribution cable.

Up to 35 mm² aluminium electricity company service cable

15 m

Fault level based on table in Electricity Association publication with a service cable 15 m long is 10.2 kA at 0.8 p.f. 400 V.

Footpath

Road

Fault level based on table in Electricity Association publication with a service cable 4 m long is 11.7 kA at 0.78 p.f. 230 V single-phase 100 A supply.

4 m

College supplied directly from electricity company substation

Fault level based on table in Electricity Association publication with a service cable 5 m long is 23.1 kA at 0.4 p.f. 400 V three-phase supply. This is based on a fault level of 25 kA at the point of connection to the electricity distributor's substation.

5 m

Dwelling

Electricity company service cable

For the declared fault level at the point of connection of the electricity supplier's substation consult the electricity distributor.

Substation

In the event that the service cable is supplied from a distribution cable in the footpath on the far side of the road the attenuation in fault level can only be applied from the footpath on the near side to the property. This is because the electricity distributor may at some time in the future install a distribution cable in the footpath nearest to the property from which to supply the service cable.

Footpath

Road

Footpath

6.2.4 Enquiry – maximum earth fault loop impedances

On enquiry electricity distributors are likely to advise that, for a PME supply, the maximum fault level is 16 000 A and the maximum earth fault loop impedance is 0.35 Ω. These two values do not seem compatible. If an open-circuit supply voltage of 250 V and a loop impedance of 0.35 Ω are assumed, a fault current of 714 A is calculated, considerably smaller than the figure for maximum fault current provided. In a similar way that a supply company will quote the maximum fault level that is likely to arise, it will also quote the highest earth fault loop impedance, since fault levels and earth fault loop impedances of the distribution network may change during the lifetime of an installation. When selecting switchgear ratings the highest fault levels must be presumed, but when using knowledge of the supply company's loop impedance for determining protection against electric shock, or the low fault current withstand of electric cables, then the highest loop impedance must be presumed.

▼ **Figure 6.2** Typical external earth fault loop impedances.
Note: values for a particular installation are obtained from the distributor.

EXTERNAL EARTH FAULT LOOP IMPEDANCE FOR 230 V SINGLE-PHASE AND 400 V THREE-PHASE SUPPLIES

A Protective Multiple Earth (PME) Terminal
(Higher values could apply to customers supplied from small capacity pole-mounted transformers and/or long lengths of low voltage overhead lines.)

Capacity up to 100 A 230 V single-phase	0.35 Ω
Capacity up to 200 A/phase 400 V three-phase	0.35 Ω
Capacity from 200 A to 300 A/phase 400 V three-phase	0.20 Ω
Capacity exceeding 300 A/phase 400 V three-phase	0.15 Ω

B Protective Neutral Bonding (PNB) Earthing Terminal
(Higher values could apply to customers supplied from small capacity pole-mounted transformers and/or long lengths of low voltage overhead lines.)

Capacity up to 100 A 230 V single-phase	0.35 Ω
Capacity up to 200 A/phase 400 V three-phase	0.35 Ω
Capacity from 200 A to 300 A/phase 400 three-phase	0.20 Ω
Capacity exceeding 300 A/phase 400 V three-phase	0.15 Ω

C Cable Sheath/Continuous Earth Wire Terminal (SNE)

Capacity up to 100 A 230 V single-phase	0.80 Ω
Capacity up to 100 A 400 V three-phase	0.80 Ω
Capacity from 100 A to 200 A/phase 400 V three-phase	0.35 Ω
Capacity from 200 A to 300 A/phase 400 V three-phase	0.20 Ω
Capacity exceeding 300 A/phase 400 V three-phase	0.15 Ω

D No Earthing Terminal provided but the Company's distribution system multiple earthed
(This external earth fault loop impedance consists of the resistance of the neutral to earth plus the impedance of the transformer winding and line conductor, but does not include the resistance of the customer's earth electrode.)

Capacity up to 100 A 230 V single-phase	21.0 Ω
All 400 V three-phase supplies	21.0 Ω

Notes:
1 For a non-PME distributor UK external fault loop impedance could be much higher.
2 Whilst these values may at first seem high, experience has shown that the more onerous restraint is likely to be due to voltage drop considerations within the installation and not external earth loop impedance.
3 PNB applies where a distribution transformer is dedicated to the supply of a single customer.

Values of maximum earth loop impedance provided by electricity companies are typically as follows for single-phase supplies:

i   TN-C-S supplies: 0.35 Ω
ii TN-S supplies: 0.8 Ω
iii TT supplies: 21 Ω

It is recommended that these values of loop impedance be used for shock protection calculations, and for assessment of the fault-carrying capability of cables (particularly reduced section protective conductors) in smaller installations up to 100 A single-phase, e.g. domestic.

(i) and (ii) above for TN-C-S and TN-S supplies are loop impedances at the origin of the installation. (iii) for TT installations is effectively the resistance of the source transformer earth electrode, and does not allow for the installation earth electrode resistance.

6.3   Determined by calculation

6.3.1   Calculation – maximum prospective fault current I_{pf}

To calculate the maximum prospective fault current I_{pf}, information on the distribution system back to the distribution transformer is required. Calculation would only be used at the design stage of a large installation, see Figure 6.3.

▼ **Figure 6.3**   Maximum and minimum prospective fault currents

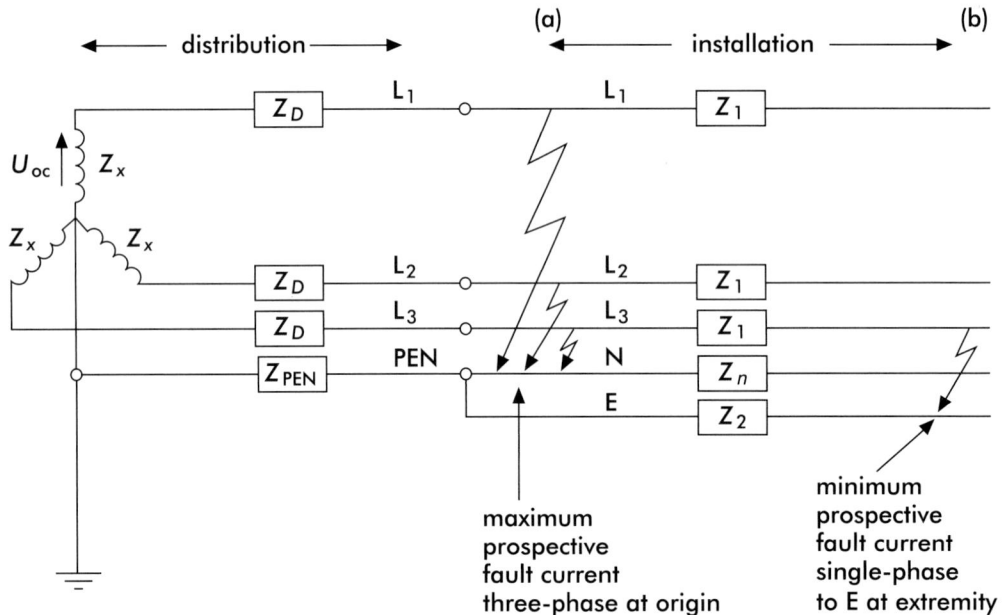

where:

U_{oc} is the open circuit phase voltage
Z_x is the phase impedance of the transformer or supply
Z_D is the line impedance of the distribution cable
Z_{PEN} is the impedance of the PEN conductor
Z_1 is the line impedance of the circuit line conductor
Z_n is the neutral impedance of the circuit neutral conductor
Z_2 is the impedance of the circuit protective conductor.

A fault across the three phases is considered the worst case, as for such a fault the neutral/earth return has no significance and in effect reduces the loop impedance, thus increasing the fault current:

$$I_{pf} = \frac{U_{oc}}{Z_x + Z_D} \cong \frac{C_{max} U_o}{Z_x + Z_D}$$

where:

I_{pf} is the maximum prospective fault current. I_{pf} is maximum when the conductors are cold, that is at ambient temperature.

C_{max} is a voltage factor for maximum fault currents, specified in BS EN 60909 for calculation purposes as 1.1 (for systems with a voltage range of +10%).

6.3.2 Calculation – minimum prospective fault current I_{ef}

To determine if a fuse or circuit-breaker will operate in the required disconnection time, the worst case, that is the situation where the fault current is at its minimum, must be considered.

The minimum prospective fault current is given by considering a single-phase to earth fault at the extremity of the circuit, see Figure 6.3:

$$I_{ef} = \frac{C_{min} U_o}{Z_x + C_r Z_D + C_r' Z_1 + C_r'' Z_2 + C_r''' Z_{PEN}}$$

where:

I_{ef} is the minimum prospective earth fault current

C_{min} is a voltage factor for minimum fault currents, specified in BS EN 60909 as 0.95

C_r etc. is a correction factor appropriate to the normal operating temperature of the conductor.

I_{ef} is calculated with the conductors of the loop at normal operating temperature, so a correction factor is applied to the resistive element of the conductors.

6.3.3 Example calculations

Figure 6.4 shows a distribution system for which the designer needs to determine

i the maximum prospective fault current at the busbars of each distribution board in order to select equipment (switch and fuse gear) of sufficient fault rating, and

ii maximum earth loop fault impedance at the end of each distribution and final circuit cable so that disconnection occurs in the event of a fault to earth in the required time. The equipment impedances are given in the tables of Appendix F.

The calculation is more easily tabulated if the resistance at operating temperature is obtained by adding a correction to the resistance at ambient. The resistance at ambient is required for the maximum short-circuit current at each board. These additive correction factors are tabulated in Appendix F, Table F.17.

The product of these correction factors and the conductor resistance at 20 °C gives the increase in resistance.

▼ Figure 6.4
Typical distribution
system

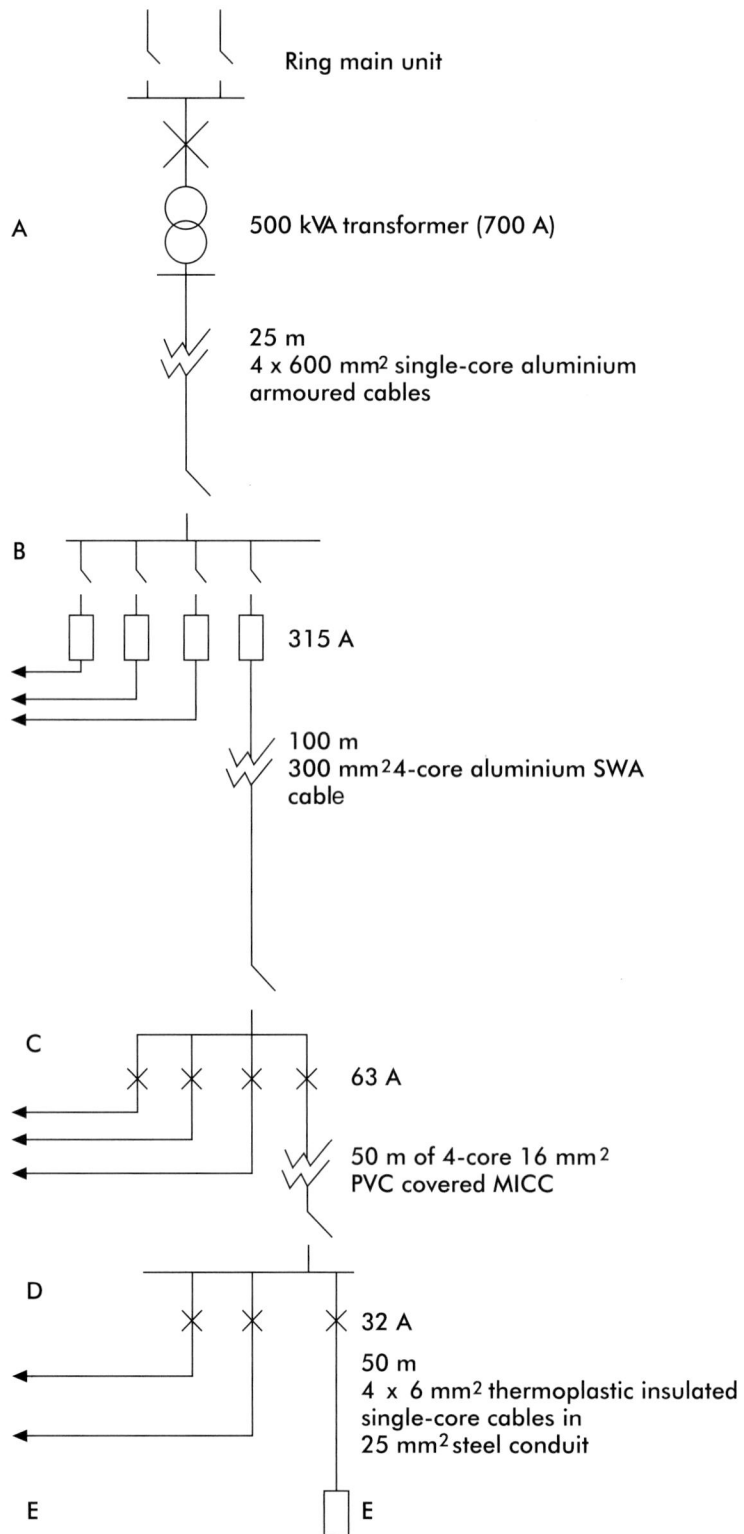

Ring main unit

A   500 kVA transformer (700 A)

25 m
4 x 600 mm² single-core aluminium
armoured cables

B

315 A

100 m
300 mm² 4-core aluminium SWA
cable

C   63 A

50 m of 4-core 16 mm²
PVC covered MICC

D   32 A

50 m
4 x 6 mm² thermoplastic insulated
single-core cables in
25 mm² steel conduit

E   E

See below for calculation of prospective fault currents for a distribution system as Figure 6.4. Impedances are taken from the tables of Appendix F.

1. Three-phase to earth fault current at B

	Impedance	
	r (Ω)	x (Ω)
500 kVA transformer (Table F.5)	0.0051	0.0171
25 m of 600 mm² single-core aluminium armoured (Table F.9): line impedance 25 x (0.0515 r + 0.09 x)/1000 Ω	0.0013	0.0023
Temperature correction (Note 1)	–	–
Total line impedance at B (Z_{pf})	0.0064	0.0194

Note 1: Temperature correction is not applied for three-phase to earth faults where the worst condition is a 'cold' installation.

$$Z_{pf} = \sqrt{r^2 + x^2} = \sqrt{0.0064^2 + 0.0194^2} = 0.0204 \ \Omega$$

$$I_{pf} = 1.1 \times 230/0.0204 = 12\ 402 \ A$$

(A phase voltage of 1.1 x 230 V is assumed for the open circuit voltage at the terminals of the transformer.)

Switchgear at B must have a fault interrupting capacity exceeding 12.5 kA. BS 88-2 fuses are to be used which have a rated short-circuit capacity of 50 kA, see Table 6.1.

2. Three-phase to earth fault current at C

	Impedance	
	r (Ω)	x (Ω)
Impedance at B from 1 above	0.0064	0.0194
100 m of 300 mm² 4-core aluminium SWA thermoplastic (Table F.7A): line impedance 100 x (0.100 r + 0.08 x)/1000 Ω	0.0100	0.0080
Temperature correction (Note 1)	–	–
Total line impedance at C (Z_{pf})	0.0164	0.0274

$$Z_{pf} = \sqrt{r^2 + x^2} = \sqrt{0.0164^2 + 0.0274^2} = 0.0319 \ \Omega$$

$$I_{pf} = 1.1 \times 230/0.0319 = 7\ 913 \ A$$

The circuit-breakers at C have to be suitable for a fault level of 8 000 A, when backed by a 315 A BS 88-2 fuse.

3. Earth fault loop impedance at C

	Impedance	
	r (Ω)	**x (Ω)**
Transformer	0.0051	0.0171
25 m of 600 mm² aluminium singles (Table F.9)		
(a) line impedance at 20 °C	0.0013	0.0023
(b) neutral/earth impedance	0.0013	0.0023
(c) correction to 70 °C (a + b) x 0.20 (Note 2)	0.0005	–
100 m of 300 mm² 4-core aluminium SWA		
(a) line impedance at 20 °C (Table F.7A)	0.0100	0.0080
(b) armour impedance at 20 °C (Table F.7A)		
100 x (0.52 r + 0.3 x)/1000 Ω	0.0520	0.030
(c) correction of (a) to 70 °C (x 0.20) (Note 2)	0.0020	–
(d) correction of (b) to 60 °C (x 0.18) (Note 3)	0.0094	–
Total earth loop impedance at C (Z_{ef})	0.0816	0.0597

Note 2: Correction factor from 20 °C to 70 °C for an aluminium conductor is (70 – 20) x 0.004 = 0.20 (see Table F.4). The correction factor is only applied to the resistive component of the impedance.

Note 3: Correction factor from 20 °C to 60 °C for cable armour (see Table 53.4 of BS 7671) is (60 – 20) x 0.0045 = 0.18 (see Table F.4).

$$Z_{ef} = \sqrt{0.0816^2 + 0.0597^2} = 0.1 \ \Omega$$

$$I_{ef} = 1.1 \times 230/0.1 = 2\,530 \text{ A}$$

From Table F.19, this fault current is sufficient to operate the 315 A fuse at B within 5 s.

4. Three-phase to earth fault current at D

	Impedance	
	r (Ω)	**x (Ω)**
Impedance at C from 2	0.0164	0.0271
50 m of 4-core 16 mm² from Table F.15 MICC 50 x 1.16r/1000 at 20 °C	0.0580	–
Temperature correction (Note 1)	–	–
Total line impedance at D (Z_{pf})	0.0744	0.0271

$$Z_{pf} = \sqrt{0.0744^2 + 0.0271^2} = 0.079 \ \Omega$$

$$I_{pf} = 1.1 \times 230/0.079 = 3\,202 \text{ A}$$

5. Earth fault loop impedance at D

	Impedance	
	r (Ω)	x (Ω)
Earth fault loop impedance at C from 3	0.0816	0.0597
50 m of 4-core 16 mm², loop impedance at 70 °C 50 x (1.4 + 0.604)r/1000 (Table F.15) (Note 4)	0.1002	–
Total earth fault loop impedance at D (Z_{ef})	0.1818	0.0597

Note 4: As the coefficient of resistance of MICC sheaths and cores is different, loop impedances are tabulated at full load temperature.

$$Z_{ef} = \sqrt{0.1818^2 + 0.0597^2} = 0.1914 \ \Omega$$

Table 41.3 of BS 7671 indicates that this loop impedance is sufficiently low for the instantaneous operation of 63 A devices at board C, types 1, 2, 3, B and C, but not D.

6. Earth fault loop impedance at E

	Impedance	
	r (Ω)	x (Ω)
Loop impedance at D from 5	0.1818	0.0597
50 m of 6 mm² thermoplastic from Table F.1 50 x 3.08r/1000	0.1540	–
Correction to 70 °C 0.154 x (70 – 20) x 0.004 (Table F.4)	0.0308	
50 m of 25 mm² steel conduit Table F.11 50 x (1.6r + 1.6x)/1000	0.0800	0.0800
Temperature correction of conduit (Note 5)	–	–
Total earth fault loop impedance at E (Z_{ef})	0.4466	0.1397

Note 5: The cross-sectional area and surface area of steel conduit is such that no increase in resistance is presumed.

$$Z_{ef} = \sqrt{0.4466^2 + 0.1397^2} = 0.4679 \ \Omega$$

All 32 A circuit-breaker types except D (and 4) operate instantaneously with a loop impedance of 0.47 Ω.

6

Shock protection   7

- ■ **Shock protection**
- ■ **Automatic disconnection of supply**

- ■ **Circuit calculations**

410 ## 7.1   Shock protection

7.1.1   Introduction

BS 7671 requires two lines of defence (protective provisions) against electric shock:

i   basic protection (against direct contact), that is, one should not be able to directly touch live parts,

ii   fault protection (formerly called protection against indirect contact).

▼ **Figure 7.1**
Basic and fault protection

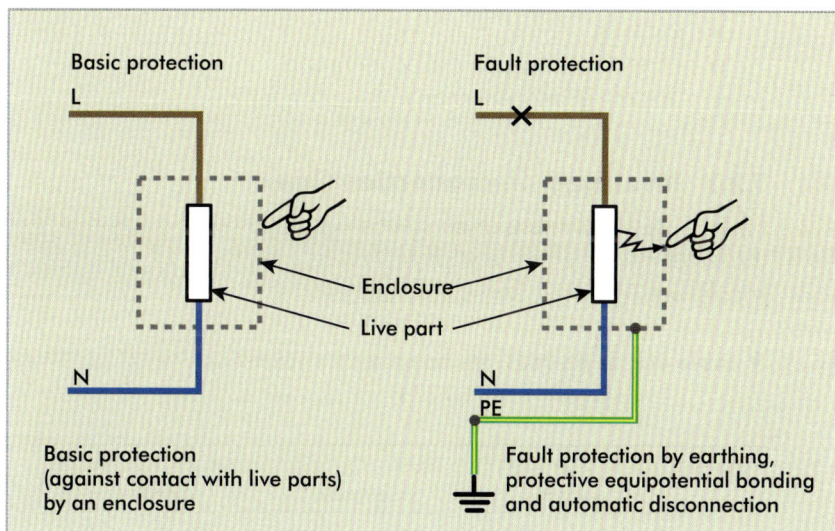

7.1.2   Protective measures

The combination of a protective provision providing basic protection and a protective provision providing fault protection is called a protective measure. Table 7.1 lists the more common protective measures.

▼ **Table 7.1** Protective measures

Protective measure	Protective provisions	
	Basic protection	**Fault protection**
Automatic disconnection of supply (411)	Insulation of live parts, barriers or enclosures	Protective earthing Automatic disconnection Protective bonding
Double insulation (412)	Basic insulation	Supplementary insulation
Reinforced insulation (412)	Reinforced insulation	
Electrical separation (413)	Insulation of live parts	Simple separation from other circuits and earth
Extra-low voltage (414)	Limitation of voltage Protective separation Basic insulation	

7.2 Protective measure: automatic disconnection of supply

The protective measure automatic disconnection of supply requires

i basic protection provided by insulation of live parts or by barriers or enclosures, and
ii fault protection provided by
  a earthing,
  b protective equipotential bonding and
  c automatic disconnection in case of a fault.

411 ### 7.2.1 Maximum disconnection times

BS 7671 sets maximum disconnection times for earth faults that, if met, will result in the circuit meeting the fault protection requirements for automatic disconnection of supply, see Table 41.1 below from BS 7671.

Table 41.1 ▼ **Table 41.1 of BS 7671** Maximum disconnection times for TN and TT systems (see Regulation 411.3.2.2)

System	$50 \, V < U_0 \leq 120 \, V$ seconds		$120 \, V < U_0 \leq 230 \, V$ seconds		$230 \, V < U_0 \leq 400 \, V$ seconds		$U_0 > 400 \, V$ seconds	
	a.c.	d.c.	a.c.	d.c.	a.c.	d.c.	a.c.	d.c.
TN	0.8	Note	0.4	5	0.2	0.4	0.1	0.1
TT	0.3	Note	0.2	0.4	0.07	0.2	0.04	0.1

Where, in a TT system, disconnection is achieved by an overcurrent protective device and protective equipotential bonding is connected to all the extraneous-conductive-parts within the installation in accordance with Regulation 411.3.1.2, the maximum disconnection times applicable to a TN system may be used.

U_0 is the nominal a.c. rms or d.c. line voltage to Earth.

Where compliance with this regulation is provided by an RCD, the disconnection times in accordance with Table 41.1 relate to prospective residual fault currents significantly higher than the rated residual operating current of the RCD.

Note: Disconnection is not required for protection against electric shock but may be required for other reasons, such as protection against thermal effects.

BS 7671 relaxes the disconnection time for TN systems to 5 seconds and for TT systems to 1 second for

a final circuits exceeding 32 A and
b distribution circuits.

7.2.2 Current causing automatic operation of protective device within the required time (I_a)

In order to achieve disconnection in the required time, it is necessary for sufficient fault current I_f to flow to rupture the fuse or operate the circuit-breaker.

When I_f is sufficiently large to cause operation of the device in the required time, it is called I_a. This current to cause disconnection in the required time is found from the device characteristics. For example see Figure 3A4 from Appendix 3 of BS 7671, reproduced below.

To avoid errors in reading off from the logarithmic axes, the key information is tabulated at the top right of the characteristics.

For example, for a 32 A type B circuit-breaker to BS EN 60898, I_a, the current to cause operation in 0.1 to 5 s is given as 160 A.

Note: For the circuit-breakers listed in BS 7671 the current I_a deemed to cause operation in 0.4 s and 5 s is the same. From the curve, a current of 159 A would result in a 32 A breaker tripping in say 15 s, whilst a current of 160 A would cause operation in 0.1 s.

▼ **Figure 3A4 of BS 7671** Time/current characteristics for type B circuit-breakers to BS EN 60898 and RCBOs to BS EN 61009-1

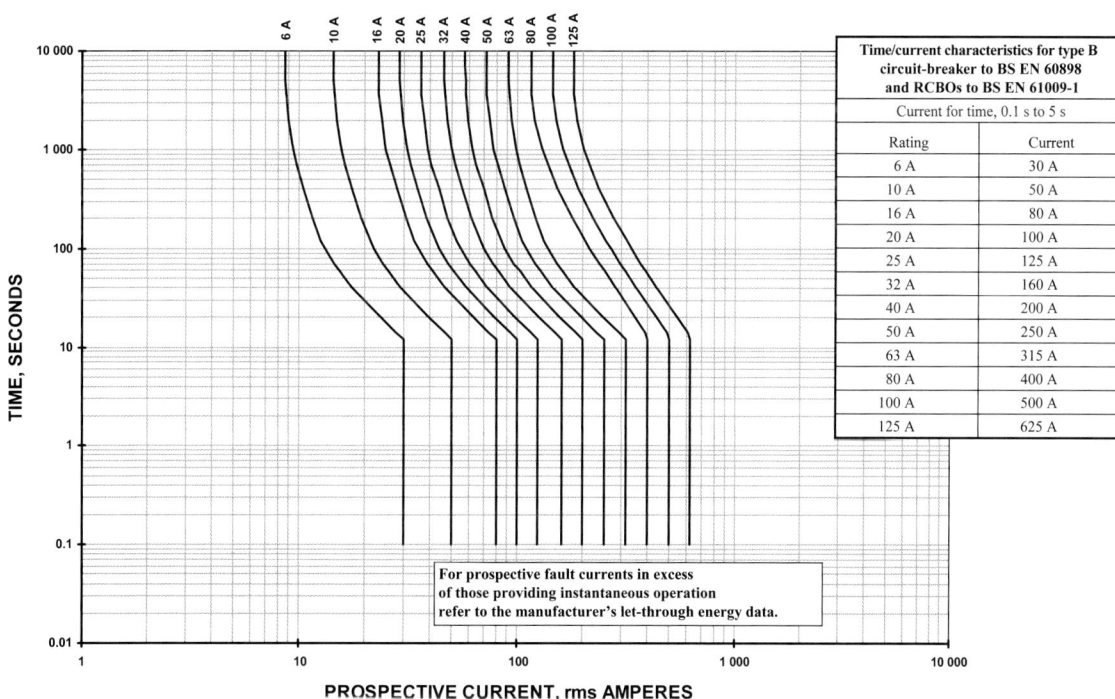

Time/current characteristics for type B circuit-breaker to BS EN 60898 and RCBOs to BS EN 61009-1	
Current for time, 0.1 s to 5 s	
Rating	Current
6 A	30 A
10 A	50 A
16 A	80 A
20 A	100 A
25 A	125 A
32 A	160 A
40 A	200 A
50 A	250 A
63 A	315 A
80 A	400 A
100 A	500 A
125 A	625 A

For prospective fault currents in excess of those providing instantaneous operation refer to the manufacturer's let-through energy data.

7.2.3 Maximum earth fault loop impedance (Z_{41})

In order for the minimum earth fault current (I_a) to flow, the earth fault loop impedance (Z_s) at the end of the circuit must not exceed the value given by:

$$Z_s \leq \frac{U_0}{I_a}$$

Example

Consider a 32 A type B circuit-breaker, from Figure 3A4 of BS 7671 above, see table to right of the curves; a current I_a of 160 A is required.

The maximum loop impedance

$$Z_s = \frac{U_0}{I_a} = \frac{230}{160} = 1.438 \ \Omega$$

Rounded up, this is the value given in Table 41.3 of BS 7671 reproduced below, i.e. 1.44 Ω.

▼ **Table 41.3 of BS 7671** Maximum earth fault loop impedance (Z_s) for circuit-breakers with U_0 of 230 V, for instantaneous operation giving compliance with the 0.4 s disconnection time of Regulation 411.3.2.2 and 5 s disconnection time of Regulation 411.3.2.3. (For RCBOs see also Regulation 411.4.9)

(a) Type B circuit-breakers to BS EN 60898 and the overcurrent characteristics of RCBOs to BS EN 61009-1

Rating (amperes)	3	6	10	16	20	25	32	40	50	63	80	100	125	I_n
Z_s (ohms)		7.67		2.87		1.84		1.15		0.73		0.46		46/I_n
	15.33		4.60		2.30		1.44		0.92		0.57		0.37	

(b) Type C circuit-breakers to BS EN 60898 and the overcurrent characteristics of RCBOs to BS EN 61009-1

Rating (amperes)	6	10	16	20	25	32	40	50	63	80	100	125	I_n
Z_s (ohms)	3.83		1.44		0.92		0.57		0.36		0.23		23/I_n
		2.30		1.15		0.72		0.46		0.29		0.18	

(c) Type D circuit-breakers to BS EN 60898 and the overcurrent characteristics of RCBOs to BS EN 61009-1

Rating (amperes)	6	10	16	20	25	32	40	50	63	80	100	125	I_n
Z_s (ohms)	1.92		0.72		0.46		0.29		0.18		0.11		11.5/
		1.15		0.57		0.36		0.23		0.14		0.09	I_n

Note: The circuit loop impedances given in the table should not be exceeded when the conductors are at their normal operating temperature. If the conductors are at a different temperature when tested, the reading should be adjusted accordingly.

These maximum values of Z_s (called Z_{41} in this guide), as the note below the table states, are not to be exceeded when the conductors are at their normal operating temperature, e.g. 70 °C for thermoplastic (PVC) insulated cables.

7.3 Circuit calculations

Circuits are designed to meet the shock protection requirements by limiting the earth fault loop impedances to the end of the circuit (Z_s) to the maximum values given in Tables 41.2 to 41.4 of BS 7671(Z_{41}).

▼ **Figure 7.2**
Simple system

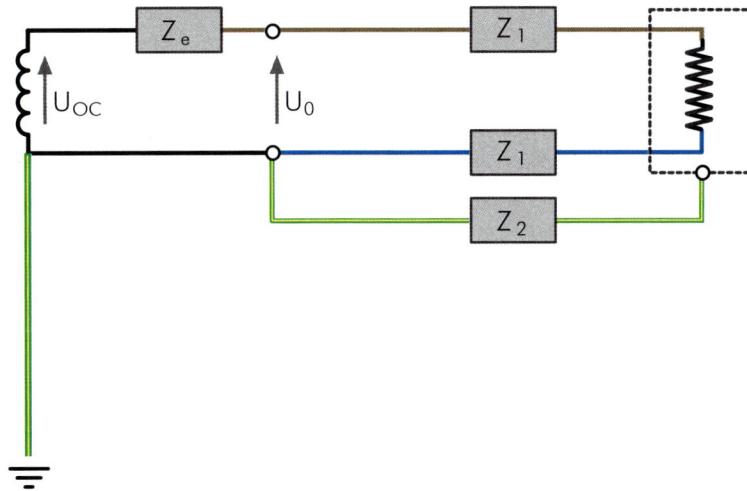

$Z_s \leq Z_{41}$ and the vector $Z_s = \dot{Z}_e + \dot{Z}_1 + \dot{Z}_2$

Hence, the vector sum $\dot{Z}_e + \dot{Z}_1 + \dot{Z}_2 \leq Z_{41}$

where:

Z_{41} is the maximum earth fault loop impedance given by the appropriate Table, 41.2, 41.3 or 41.4

Z_e is the earth fault loop impedance external to the circuit – in this section it is assumed to be that of the supply, i.e. 0.8 or 0.35 Ω

Z_1 is the impedance of the line conductor of the circuit

Z_2 is the impedance of the protective conductor of the circuit.

Values of Z''_1 and Z''_2 per metre for common cables and equipment are given in Appendix F in ohms per metre at 20 °C.

For the purposes of shock protection calculations these will need to be multiplied by the circuit length L and corrected to the conductor operating temperature, say 70 °C. The correction factor C_r is applied only to the resistive element.

The correction factors C_r are given in Table F.3 in Appendix F.

The equation for fault protection against electric shock becomes (for radial circuits):

the vector sum $\dot{Z}_e + \left(R''_1 + R''_2\right) \times C_r \times L + \left(X''_1 + X''_2\right) \times L \leq Z_{41}$

Arithmetically, $Z_e + \sqrt{(R''_1 + R''_2)^2 \times C_r^2 \times L^2 + (X''_1 + X''_2)^2 \times L^2} \leq Z_{41}$

For circuits with a conductor size 16 mm^2 or less, the inductance X is not significant and the equation becomes $Z_e + (R''_1 + R''_2) \times C_r \times L \leq Z_{41}$

where:

> R''_1 is the resistance per metre of the line conductor (see Tables F.1 and F.7)
> R''_2 is the resistance per metre of the protective conductor (see Tables F.1 and F.7)
> X''_1 is the reactance per metre of the line conductor (see Table F.7)
> X''_2 is the reactance per metre of the protective conductor (see Table F.7)
> C_r is the correction factor for temperature (see Table F.3)
> L is the length of cable in the circuit.

The maximum circuit cable lengths, L_s, that will limit the circuit resistance such that disconnection in the event of a fault to earth occurs within the required time (0.4 s or 5 s) is given by:

For radial circuits: $L_s = \dfrac{Z_{41} - Z_e}{(R''_1 + R''_2) C_r}$

For ring circuits see Chapter 2 and Appendix B.

Example

Consider a 9.6 kW shower circuit protected by a 40 A type C circuit-breaker with a PME supply, to be wired in 10 mm^2 thermoplastic (PVC) insulated and sheathed flat cable with 4 mm^2 protective conductor (Table 4D5 of BS 7671). What is the maximum length L_s to meet the shock protection requirement?

Z_{41} from Table 41.3(b) is 0.57 Ω. Maximum Z_e for PME supply is 0.35 Ω.

$R''_1 + R''_2 = 6.44$ mΩ/m at 20 °C from Table F.1. C_r from Table F.3 is 1.20.

Hence $L_s = \dfrac{Z_{41} - Z_e}{(R''_1 + R''_2) C_r} = \dfrac{0.57 - 0.35}{\left(\dfrac{6.44}{1000}\right)1.20} = 28.5$ m.

For full designs where reactance is not ignored and where protective conductors may be cable armouring or similar, it assists in the calculation if the correction factor is presented in a slightly different way, see Table F.17 and section 6.3.2.

Protection against fault current

8

- The adiabatic equation
- Selection from Table 54.7
- Introduction to calculations
- Simple calculation
- Energy let-through
- Plotting adiabatics

- Sheath or armour of a cable
- Plotting armour adiabatics
- Armour capability
- Conduit and trunking
- Earthing and bonding conductors

543
434.5.2
8.1 The adiabatic equation

543.1.3
8.1.1 Introduction to the adiabatic equation

All conductors – line, neutral and protective – must meet the size requirements of the adiabatic equation:

$$S \geq \frac{\sqrt{I^2 t}}{k}$$

where:

- S   is the nominal cross-sectional area of conductor in mm²
- I   is the value of fault current in amperes, expressed for a.c. as the rms value, due account being taken of the current-limiting effect of the circuit impedances
- t   is the duration in seconds of the fault current
- k   is a factor taking account of the resistivity, temperature coefficient and heat capacity of the conductor material, and the appropriate initial and final temperatures. For common materials the values of k are shown in Table 43.1 of BS 7671.

435.1
8.1.2 Protection by one device

Where the overcurrent device meets the requirements for overload protection the adiabatic equation is satisfied if the current-carrying capacity (I_z) of the line and neutral conductors meets the requirements of Regulation 433.1.1, that is:

$$I_z \geq I_n \geq I_b$$

where:

- I_z is the current-carrying capacity of a cable for continuous service, under the particular installation conditions concerned
- I_n is the rated current or current setting of the device protecting the circuit against overcurrent
- I_b is the design current of the circuit, i.e. the current intended to be carried by the circuit in normal service.

The adiabatic equation will also be met for neutral and protective conductors of the same current-carrying capacity as the line conductor. (The term protective conductor includes circuit protective conductor, earthing conductor and bonding conductor.) However, if the protective (or neutral) conductors are of lesser current-carrying capacity, for example because they have reduced cross-sectional area or are of a different material or construction, they must:

i   be selected from Table 54.7, see section 8.2, or
ii comply with the adiabatic equation, see section 8.1.1.

434 8.1.3 Fault current protection

If the circuit has no overload protection (that is, only fault protection) compliance with the adiabatic equation must be confirmed for the line and neutral conductors. The protective conductor must be selected from Table 54.7 or comply with the adiabatic equation as for overload protection.

8.1.4 Selection or calculation

The simplest way of checking compliance of a reduced current-carrying capacity protective conductor is to confirm the protective conductor meets Table 54.7.

However, it is common to use reduced section protective conductors particularly for final circuits and these do not comply with Table 54.7, e.g. socket-outlet circuits wired in twin and cpc cable with live conductors of 2.5 mm^2 and a protective conductor of 1.5 mm^2.

A set of tables has been prepared to enable compliance to be checked where standard devices included in BS 7671 have been used, that is fuses to BS 88, BS 1361, BS 3036 and circuit-breakers to BS EN 60898 or BS EN 61009. This can be done because the standards specify sufficient performance requirements for the maximum disconnection time to be known.

For other devices the adiabatic calculation will need to be carried out with manufacturer's data. The energy let-through calculations are the simplest, if manufacturer's data is available.

543.1.4 8.2 Selection from Table 54.7

▼ **Table 54.7 of BS 7671** Minimum cross-sectional area of protective conductor in relation to the cross-sectional area of associated line conductor

Cross-sectional area of line conductor, S	Minimum cross-sectional area of the corresponding protective conductor	
	If the protective conductor is of the same material as the line conductor	If the protective conductor is not the same material as the line conductor
(mm²)	(mm²)	(mm²)
$S \leq 16$	S	$\dfrac{k_1}{k_2} \times S$
$16 < S \leq 35$	16	$\dfrac{k_1}{k_2} \times 16$
$S > 35$	$\dfrac{S}{2}$	$\dfrac{k_1}{k_2} \times \dfrac{S}{2}$

where:

k_1 is the value of k for the line conductor, selected from Table 43.1 in Chapter 43 according to the materials of both conductor and insulation.

k_2 is the value of k for the protective conductor, selected from Tables 54.2 to 54.6 as applicable.

Protective conductors of same material as line and neutral conductors

Here, a simple comparison is made with column 2 above.

For example, if wiring a circuit with 10 mm² live conductors in plastic conduit the protective conductor would also need to be 10 mm² to comply with Table 54.7.

A 35 mm² line conductor requires a 16 mm² protective conductor and a 120 mm² line conductor a 70 mm² protective conductor (next standard size up from 60 mm²).

Protective conductors of a different material to the line and neutral conductors

The third column is used and requires the 'k' values for the line and neutral conductors and for the protective conductor. The values of k for protective conductors are given in Tables 54.2 to 54.6 of BS 7671, reproduced here as consolidated Table 8.1.

The tables are used as follows:

Protective conductor type	Table for k
Separate protective conductor on own, not in trunking	54.2
Conductor as a core of a cable or bunched	54.3
Cable in conduit or trunking	54.3
Sheath or armour of cable	54.4
Conduit or trunking as protective conductor	54.5
Bare conductor (see table heading)	54.6

▼ **Table 8.1**   Values of k from Tables 54.2 to 54.6 of BS 7671

TABLE 54.2
Values of k for insulated protective conductor not incorporated in a cable and not bunched with cables where the assumed initial temperature is 30 °C

Material of conductor	Insulation of protective conductor or cable covering		
	70 °C thermoplastic	90 °C thermoplastic	90 °C thermosetting
Copper	143/133*	143/133*	176
Aluminium	95/88*	95/88*	116
Steel	52	52	64
Assumed initial temperature	30°C	30 °C	30 °C
Final temperature	160 °C/140 °C*	160 °C/140 °C*	250 °C

TABLE 54.3
Values of k for protective conductor incorporated in a cable or bunched with cables where the assumed initial temperature is 70 °C or greater

Material of conductor	Insulation material		
	70 °C thermoplastic	90 °C thermoplastic	90 °C thermosetting
Copper	115/103*	100/86*	143
Aluminium	76/68*	66/57*	94
Assumed initial temperature	70 °C	90 °C	90 °C
Final temperature	160 °C/140 °C*	160 °C/140 °C*	250 °C

TABLE 54.4
Values of k for insulated protective conductor as a sheath or armour of a cable

Material of conductor	Insulation material		
	70 °C thermoplastic	90 °C thermoplastic	90 °C thermosetting
Aluminium	93	85	85
Steel	51	46	46
Lead	26	23	23
Assumed initial temperature	60 °C	80 °C	80 °C
Final temperature	200 °C	200 °C	200 °C

TABLE 54.5
Values of k for steel conduit, ducting and trunking as the protective conductor

Material of protective conductor conduit	Insulation material		
	70 °C thermoplastic	90 °C thermoplastic	90 °C thermosetting
Steel conduit, ducting and trunking	47	44	58
Assumed initial temperature	50 °C	60 °C	60 °C
Final temperature	160 °C	160 °C	250 °C

TABLE 54.6
Values of k for bare conductor where there is no risk of damage to any neighbouring material by the temperatures indicated. The temperatures indicated are valid only where they do not impair the quality of the connections

Material of conductor	Conditions		
	Visible and in restricted areas	Normal conditions	Fire Risk
Copper	228	159	138
Aluminium	125	105	91
Steel	82	58	50
Assumed initial temperature	30 °C	30 °C	30 °C
Final temperature			
Copper conductors	500 °C	200 °C	150 °C
Aluminium conductors	300 °C	200 °C	150 °C
Steel conductors	500 °C	200 °C	150 °C

* Above 300 mm^2

8.3　Introduction to calculations

The calculations can be carried out in three different ways:

i　individual calculation from device time/current characteristics – see section 8.4
ii　using energy let-through data – see section 8.5
iii　by plotting protective conductor adiabatic characteristics on the device time/
current characteristics – see sections 8.6 and 8.8.

For fuses (to BS 88 and BS 3036) and circuit-breakers (BS EN 60898) calculation can be avoided by reference to Tables 8.4 to 8.8 for copper conductors and 8.9 and 8.10 for steel wire armouring.

For moulded case circuit-breakers (to BS EN 60947-2) a simple energy let-through calculation is required using manufacturer's data, see section 8.5.

Protective conductors are required to carry leakage currents and earth fault currents.

The requirement to carry leakage currents generally imposes no constraints upon the size of the conductors as they will almost certainly be very small compared with the steady-state current rating (tabulated in Appendix 4 of BS 7671), other than that necessary for mechanical strength.

8.4　Simple calculation

The simple calculation is used for single calculations where the loop impedance at the end of the circuit Z_s is known and for circuit-breakers to BS EN 60898.

8.4.1　Z_s known

The fault current I_f at the end of the circuit can be calculated from:

$$I_f = \frac{U_0}{Z_s}$$

where:

U_0　is the nominal line to earth voltage
Z_s　is the earth fault loop impedance.

Then the adiabatic equation can be used:

$$S \geq \frac{\sqrt{I^2 t}}{k}$$

The disconnection time t is determined from the device characteristics, e.g. Appendix 3 of BS 7671. Factor k is obtained from Tables 54.2 to 54.6, see consolidated Table 8.1 above.

Note: If the protective conductor has the same current-carrying capability as the line and neutral conductors, and the overcurrent device is providing protection against overload and fault currents, no further checks need be carried out. However, if the equivalent csa of the protective conductor is less than that of the line and neutral conductors, compliance with the adiabatic equation must be confirmed.

For a fuse the most onerous condition for the protective conductor occurs when the fault current is lowest and disconnection time longest, see Figure 4.1a of this guide. For a circuit-breaker (see Figure 4.1b) the relationship between energy let-through and fault level is more complicated.

Example

Consider a circuit to be wired in 4/1.5 mm^2 twin with earth cable of length 30 m. The supply is TN-S and the protective device is a 32 A cartridge fuse to BS 88-3.

$Z_s = Z_e + (R_1 + R_2)$, that is $Z_s = Z_e + (R''_1 + R''_2) \times L \times C_r$

Z_e is taken to be 0.8 ohm, as TN-S (section 2.2.3).

From Table F.1 $(R''_1 + R''_2) = 16.71$ mΩ/m, from Table F.3 $C_r = 1.20$

$(R''_1 + R''_2) = 16.71/1000$ Ω/m

$Z_s = 0.8 + (16.71/1000 \times 30 \times 1.20) = 1.40$ ohms

Fault current $I_f = U_0/Z_s = 230/1.40 = 164$ A

From Figure 3A1 of BS 7671 (see below) the disconnection time is approximately 2 s.

▼ **Figure 3A1 of BS 7671** Time/current characteristics for fuses to BS 88-3 fuse system C

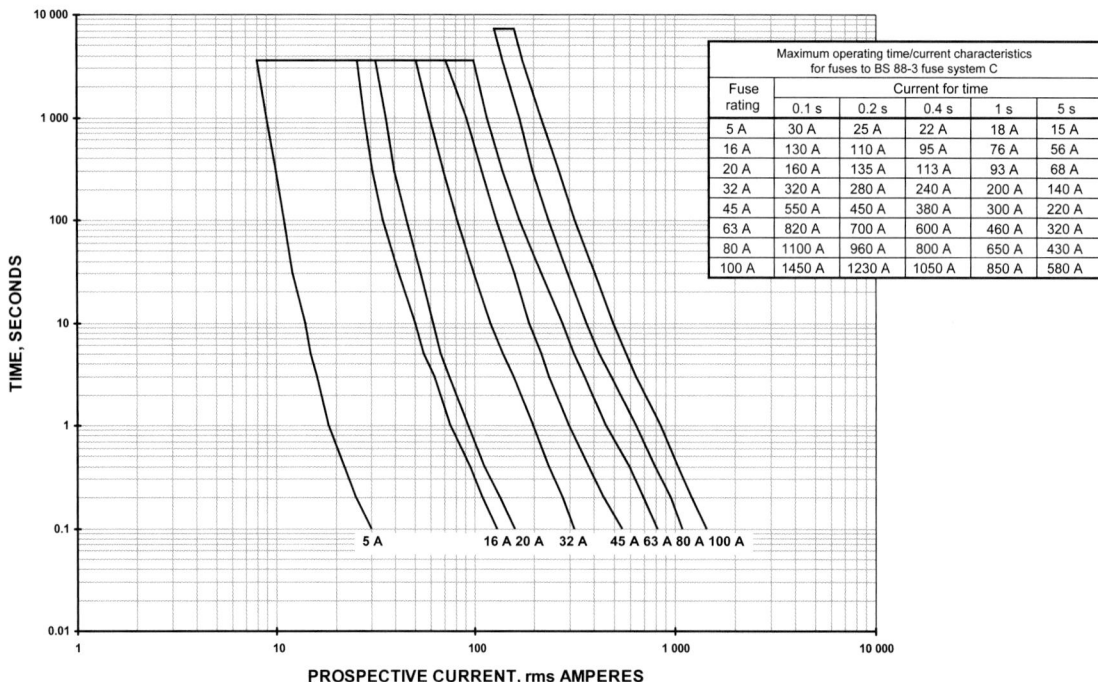

Fuse rating	Current for time				
	0.1 s	0.2 s	0.4 s	1 s	5 s
5 A	30 A	25 A	22 A	18 A	15 A
16 A	130 A	110 A	95 A	76 A	56 A
20 A	160 A	135 A	113 A	93 A	68 A
32 A	320 A	280 A	240 A	200 A	140 A
45 A	550 A	450 A	380 A	300 A	220 A
63 A	820 A	700 A	600 A	460 A	320 A
80 A	1100 A	960 A	800 A	650 A	430 A
100 A	1450 A	1230 A	1050 A	850 A	580 A

Maximum operating time/current characteristics for fuses to BS 88-3 fuse system C

From Table 54.3 in Table 8.1, k = 115

The minimum csa of the conductor S is given by $S \geq \dfrac{\sqrt{I^2 t}}{k}$

$$S = \frac{\sqrt{164^2 \times 2}}{115} = 2.02 \text{ mm}^2$$

The 4.0/1.5 mm² cable would not be suitable. A 6 mm² cable with a 2.5 mm² protective conductor could be used, or single-core cables in conduit with at least a 2.5 mm² protective conductor.

8.4.2 Circuit-breakers

If it is assumed that the circuit will be designed so that there is instantaneous operation in no less than 0.1 s, that is at low fault levels, simple calculations of minimum protective conductor sizes can be carried out, see Table 8.2 for a type D circuit-breaker. A disconnection time (t) of 0.1 s and a k of 115 are assumed. The current is obtained from Figure 3A6 of BS 7671 or is 20 times the device rating.

However, at high fault levels, say 3kA and above, Regulation 434.5.2 requires a check that the energy let-through of the device does not exceed the k²S² of the cable, see section 8.5.

▼ **Table 8.2**   Calculation of minimum protective conductor csa for type D circuit-breakers to BS EN 60898 and RCBOs to BS EN 61009

Rated current	Current I for time, 0.1 s to 5 s	$S_{min} = \frac{\sqrt{I^2 t}}{k}$ *	Practical minimum size, S_a
(A)	(A)	(mm²)	(mm²)
6	120	0.33	1.0
10	200	0.50	1.0
16	320	0.87	1.0
20	400	1.09	1.5
25	500	1.37	1.5
32	640	1.75	2.5
40	800	2.19	2.5
50	1000	2.74	4.0
63	1260	3.46	4.0
80	1600	4.39	6.0
100	2000	5.49	6.0

* The instantaneous disconnection time is assumed to be 0.1 s, and k is assumed to be 115, see Table 8.1 (Table 54.3).

8.5   Energy let-through calculation

The energy let through by a circuit-breaker determines the minimum size of the downstream cable. The cable withstand depends upon the conductor material, the insulation used and the conductor size. Manufacturers of circuit-breakers provide energy let-through data for their devices, which allow the adequacy of protective conductors to be checked. The manufacturer's data must be used for the particular device (frame size and rating) as they differ sufficiently from manufacturer to manufacturer to prevent standard tables from being prepared. Data will be presented in a similar manner to Table 8.3.

▼ **Table 8.3**   Manufacturer's energy let-through (I^2t) data for circuit-breakers

Frame	Ratings A	Maximum let-through at various prospective fault currents (amperes^2 seconds x 10^6)						
		10 kA	20 kA	25 kA	30 kA	36 kA	40 kA	50 kA
CD	16–100	0.28	0.42	0.47				
CN	125–250	0.52	0.70	0.71	0.72	0.73		
CH	16–250	0.52	0.70	0.71	0.72	0.73	0.74	0.75
SMA	300–800		8.6	12	15	21	25	36

For thermoplastic insulated cables the thermal withstand in amperes^2 seconds x 10^6 (k = 115)

Cable size (mm^2)	4	6	10	16	25	35	50
Maximum thermal stress I^2t	0.212	0.476	1.32	3.4	8.26	16.2	33.1

Using the formula $S \geq \dfrac{\sqrt{I^2 t}}{k}$ and looking up k in Tables 54.2 to 54.6 (Table 8.1), S the minimum cross-sectional area can be calculated.

Example

Consider a 200 A circuit-breaker (frame types CN or CH above) installed in a location where the fault level is 20 kA protecting a 95 mm^2 4-core armoured thermoplastic insulated cable. Is the armouring of sufficient size?

From the table above, the I^2t energy let-through would be 0.7 x 10^6 ampere^2 seconds.

From Table 54.4 in Table 8.1 above, k is 51

and $S = \dfrac{\sqrt{I^2 t}}{k} = \dfrac{\sqrt{0.7 \times 10^6}}{51} = 16.4 \text{ mm}^2$

From Table F.7B, area of the cable armouring is 160 mm^2, which is clearly sufficient.

8.5.1   Types B, C and D circuit-breakers and RCBOs

Regulation 434.5.2 requires that for faults of very short duration (i.e. at high fault levels) the conductors of the circuit shall be able to withstand the let-through energy of the protective device, i.e.

$$k^2 S^2 \geq \text{the let-through energy of the device } (I^2 t)$$

where:

S = nominal cross-sectional area of the conductor in mm^2
k = factor from Table 43.1
I^2t = the energy let-through quoted for the class of device in BS EN 60898-1, BS EN 61009-1, or as quoted by the manufacturer.

The energy let-through quoted for the class of device in BS EN 60898-1 and BS EN 61009-1 is reproduced in Table 8.4 and the minimum value of S calculated for a range of fault levels (prospective fault currents).

▼ **Table 8.4** Energy limiting class 3 Types B and C circuit-breakers and RCBOs, minimum copper protective conductor sizes determined from maximum energy let-through allowed by annex ZA of BS EN 60898-1 and annex ZD of BS EN 61009-1 for k =115

Device rating	Prospective fault current	Type B			Type C		
		Maximum energy let-through I^2t	Minimum copper protective conductor csa	Selected copper protective conductor	Maximum energy let-through I^2t	Minimum copper protective conductor csa	Selected copper protective conductor
(A)	(A)	(A²s)	S* (mm²)	(mm²)	(A²s)	S* (mm²)	(mm²)
3 to 16	3 000	15 000	1.06	1.0	18 000	1.16	1.5
	4 500	25 000	1.4	1.5	30 000	1.51	1.5
	6 000	35 000	1.6	2.5	42 000	1.78	2.5
	10 000	70 000	2.3	2.5	84 000	2.52	2.5
20 to 32	3 000	18 000	1.16	1.5	22 000	1.29	1.5
	4 500	32 000	1.56	1.5	39 000	1.72	2.5
	6 000	45 000	1.84	2.5	55 000	2.04	2.5
	10 000	90 000	2.61	4	110 000	2.88	4
40	3 000	21 600	1.27	1.5	26 400	1.41	1.5
	4 500	38 400	1.7	1.5	46 800	1.88	2.5
	6 000	54 000	2.02	2.5	66 000	2.23	2.5
	10 000	108 000	2.86	4	132 000	3.16	4

* Note $S \geq \dfrac{\sqrt{\text{energy let-through } (I^2t)}}{k}$ where $k = 115$

Energy let-through quoted by manufacturers is usually less than that quoted in BS EN 60898-1 and BS EN 61009-1, see Table 8.5.

▼ **Table 8.5a** Minimum values of S calculated using manufacturers' quoted energy let-through at 6 kA fault level

Rating	MK Type B			MK Type C		
	Energy let-through (kA²s)	S minimum (mm²)	Selected S (mm²)	Energy let-through (kA²s)	S minimum (mm²)	Selected S (mm²)
6	14	1.02	1	17	1.1	1.5
10			1.5			2.5
16	30	1.51	1.5	40	1.74	2.5
20			2.5			2.5
32	42	1.78	2.5	50	1.94	2.5
40	60	2.13	2.5	60	2.13	2.5
63	70	2.3	2.5	70	2.45	2.5

Rating	Hager Types B and C			Hager Type D		
	Energy let-through (kA²s)	S minimum (mm²)	Selected S (mm²)	Energy let-through (kA²s)	S minimum (mm²)	Selected S (mm²)
6	13	0.99	1	14	1.03	1.5
10	15	1.06	1.5	18	1.17	1.5
16	18	1.17	1.5	23	1.32	1.5
20	23	1.31	1.5	33	1.58	2.5
32	29	1.48	1.5	47	1.89	2.5
40	29	1.48	1.5	47	1.89	2.5
63	37	1.62	2.5	60	2.12	2.5

▼ **Table 8.5b** Minimum values of S calculated using manufacturers' quoted energy let-through at 3 kA fault level

Rating	MK Type B			MK Type C		
	Energy let-through (kA²s)	S minimum (mm²)	Selected S (mm²)	Energy let-through (kA²s)	S minimum (mm²)	Selected S (mm²)
6	6	0.67	1	7	0.7	1
10						
16	12	0.95	1	16	1.1	1.5
20						1.5
32	16	1.1	1.5	20	1.2	1.5
40	19	1.2	1.5	24	1.3	1.5
63	26	1.4	1.5	32	1.5	1.5

continues

▼ Table 8.5b *continued*

Rating	Hager Types B and C			Hager Type D		
	Energy let-through (kA²s)	S minimum (mm²)	Selected S (mm²)	Energy let-through (kA²s)	S minimum (mm²)	Selected S (mm²)
6	6.7	0.7	1	7.1	0.7	1
10	7.7	0.8	1	9	0.8	1
16	9.1	0.8	1	11	0.9	1
20	11.3	0.9	1	16	1.1	1.5
32	14.3	1.0	1	21	1.2	1.5
40	14.3	1.0	1	21	1.2	1.5
63	16	1.1	1.5	26	1.4	1.5

8.6 Plotting protective conductor adiabatics

This approach is suitable for preparing standard circuits to find a limiting case.

The suitability of a protective conductor can be determined graphically (see Figure 8.1). The thermal characteristics of the protective conductor are plotted on the device time/current characteristics. For a given size of protective conductor S, values of I are assumed and values of t then determined from the equation $S = \dfrac{\sqrt{I^2 t}}{k}$, hence $t = \dfrac{k^2 S^2}{I^2}$.

These will be the maximum disconnection times if the adiabatic equation is to be met.

The plot will be a straight line on the log/log graph paper of the device time/current characteristics. The point of intersection is the minimum fault current (or maximum loop impedance) at which the protective conductor is protected by the particular device, see Figure 8.1.

The intersection of the conductor characteristic with that of the overcurrent device gives the minimum adiabatic prospective earth fault current, I_a, required to protect the conductor. From this minimum adiabatic current (I_a), the maximum loop impedance (Z_a) allowed can be calculated

for $Z_a = \dfrac{230}{I_a}$

▼ **Figure 8.1** Conductor adiabatic plotted on the BS 88-3 fuse characteristics from Figure 3A1 of Appendix 3 of BS 7671

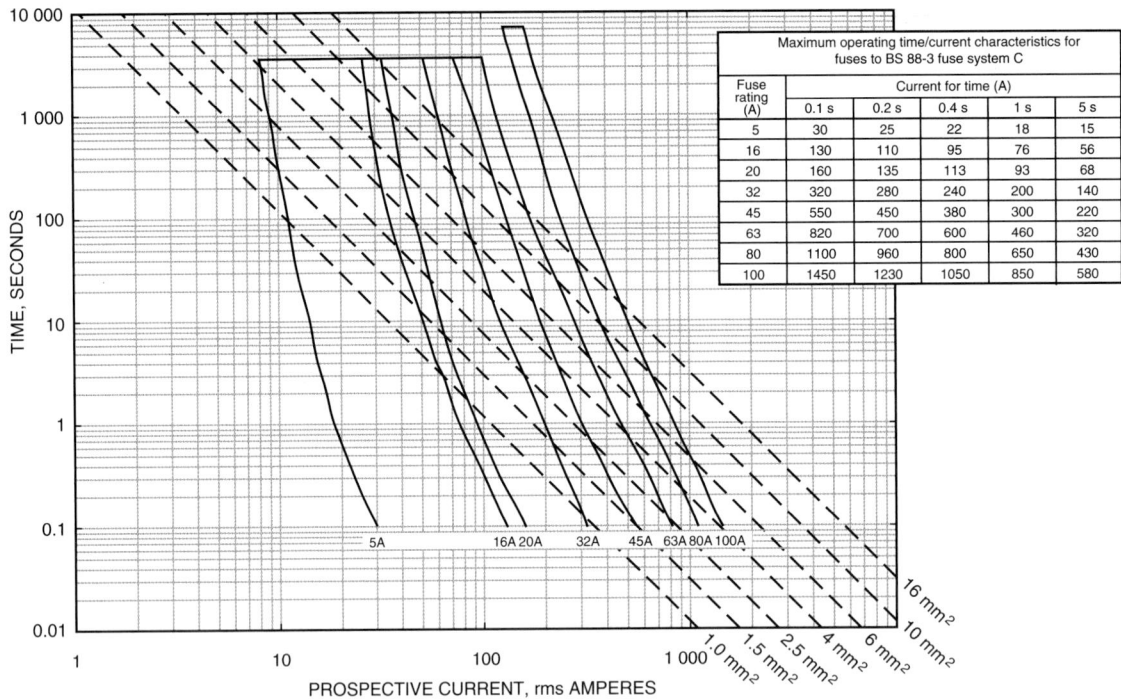

Maximum operating time/current characteristics for fuses to BS 88-3 fuse system C					
Fuse rating (A)	Current for time (A)				
	0.1 s	0.2 s	0.4 s	1 s	5 s
5	30	25	22	18	15
16	130	110	95	76	56
20	160	135	113	93	68
32	320	280	240	200	140
45	550	450	380	300	220
63	820	700	600	460	320
80	1100	960	800	650	430
100	1450	1230	1050	850	580

If we consider in Figure 8.1 the intersection of the 1.0 mm^2 conductor with each of the fuse characteristics, we can estimate the minimum adiabatic fault current (I_a) and the maximum adiabatic loop impedance (Z_a) as below:

BS 88-3 fuse rating (A)	5	16	20	32	45	63
Minimum adiabatic fault current, I_a, for 1.0 mm^2 protective conductor (A)	11	62	80	300	NP	NP
Maximum loop impedance, Z_a, at 70 °C (Ω)	21	3.7	2.9	0.8	NP	NP

NP Not permitted

These maximum adiabatic loop impedances, Z_a, can be used to calculate the maximum circuit length allowed, in the same way as Z_s is used to determine the maximum circuit length for shock protection.

Please see standard circuit calculations in Chapter 2. The corrections for temperature are identical. The maximum adiabatic loop impedances at conductor operating temperature for the fuses in BS 7671 are found in Tables 8.6, 8.7 and 8.8.

Providing a circuit-breaker is operating instantaneously, relatively small protective conductors can handle the energy let-through. Minimum protective conductor size can also be determined from energy let-through (I^2t) figures provided by manufacturers for

$$S \geq \frac{\sqrt{I^2 t}}{k} = \frac{\sqrt{\text{energy let-through}}}{k}, \text{ see section 8.5 above.}$$

▼ **Table 8.6**   Cartridge fuses to BS 88-3. Maximum adiabatic loop impedance, Z_a, for copper conductors with 70 °C thermoplastic insulation incorporated in a cable or bunched with cables at conductor operating temperature.

Protective conductor csa (mm²)	Z_a (Ω) for various device ratings							
	5 A	16 A	20 A	32 A	45 A	63 A	80 A	100 A
1.0	21	3.7*	2.9*	0.8*	NP	NP	NP	NP
1.5	‡	4.6	3.7	1.1*	NP	NP	NP	NP
2.5	‡	‡	‡	1.64	0.8*	0.3*	NP	NP
4.0	‡	‡	‡	2.1	1.1	0.54*	0.38*	0.2*
6.0	‡	‡	‡	‡	‡	0.77	0.4*	0.3*
10	‡	‡	‡	‡	‡	‡	‡	0.4
16	‡	‡	‡	‡	‡	‡	‡	‡
Z_s†	15.33	4.11	3.38	1.64	1.04	0.72	0.53	0.40

Notes:

*   loop impedance is less than that required for 5 s disconnection – see Table 41.4.

†   maximum loop impedance to meet the 5 s disconnection limit of Table 41.4.

‡   loop impedance well exceeds that required for 5 s disconnection.

NP Not permitted at any loop impedance.

▼ **Table 8.7**   Semi-enclosed fuses to BS 3036. Maximum adiabatic loop impedance, Z_a, for copper conductors with 70 °C thermoplastic insulation incorporated in a cable or bunched with cables at conductor operating temperature.

Protective conductor csa (mm²)	Z_a (Ω) for various device ratings						
	5 A	15 A	20 A	30 A	45 A	60 A	100 A
1.0	23	5.5	3.4*	NP	NP	NP	NP
1.5	24	6.3	4.3	2.5*	NP	NP	NP
2.5	‡	‡	‡	3.2	1.4*	NP	NP
4.0	‡	‡	‡	‡	1.9	1.0*	NP
6.0	‡	‡	‡	‡	‡	1.4	NP
10.0	‡	‡	‡	‡	‡	‡	0.60
Z_s†	17.7	5.35	3.83	2.64	1.59	1.12	0.53

Notes:

*   loop impedance is less than that required for 5 s disconnection – see Table 41.4.

†   maximum loop impedance to meet the 5 s disconnection limit of Table 41.4.

‡   loop impedance well exceeds that required for 5 s disconnection.

NP Not permitted at any loop impedance.

▼ **Table 8.8** Cartridge fuses to BS 88-2. Maximum adiabatic loop impedance, Z_a, for copper conductors with 70 °C thermoplastic insulation incorporated in a cable or bunched with cables at conductor operating temperature.

Protective conductor csa (mm²)	Z_a (Ω) for various device ratings							
	6 A	10 A	16 A	20 A	25 A	32 A	40 A	50 A
1.0	19	8.5	3.9*	2.1*	1.5*	0.8*	NP	NP
1.5	‡	‡	4.8	3.1	2.1*	1.3*	0.8*	NP
2.5	‡	‡	‡	‡	2.6	1.9	1.1*	0.7*
4.0	‡	‡	‡	‡	‡	2.4	1.4	1.0
6.0	‡	‡	‡	‡	‡	‡	‡	‡
Z_s†	12.8	7.19	4.18	2.95	2.30	1.84	1.35	1.04

Protective conductor csa (mm²)	Z_a (Ω) for various device ratings					
	63A	80 A	100 A	125 A	160 A	200 A
1.5	NP	NP	NP	NP	NP	NP
2.5	0.42*	NP	NP	NP	NP	NP
4.0	0.62*	0.28*	NP	NP	NP	NP
6.0	0.88	0.40*	0.27*	0.15*	NP	NP
10.0	‡	0.71	0.40*	0.25*	NP	NP
16.0	‡	‡	0.56	0.35	0.24*	0.13*
25.0	‡	‡	‡	0.46	0.35	0.24
Z_s†	0.82	0.57	0.46	0.34	0.28	0.19

Notes:

* loop impedance is less than that required for 5 s disconnection – see Table 41.4.

† maximum loop impedance to meet the 5 s disconnection limit of Table 41.4.

‡ loop impedance well exceeds that required for 5 s disconnection.

NP Not permitted at any loop impedance.

8.7 Protective conductor as a sheath or armour of a cable

The armour or sheath of a cable can be used as a protective conductor. This is a desirable arrangement as it will result in a low earth loop impedance, as the armour is in good inductive contact with the line conductors.

Regulation 543.1.4 allows Table 54.7 to be used where it is not wished to calculate the minimum cross-sectional area of a protective conductor. It will be found that there are some cables, particularly thermosetting insulated cables, which do not meet the requirements of Table 54.7. In this circumstance the adiabatic characteristics of the cable need to be considered.

Tables 8.9 and 8.10 list the minimum area of steel wire armour required by any of the BS EN 60269-2 and BS 88 fuses, and in conjunction with Tables F.7A and F.7B the adequacy of the armour of any cable can be determined. For circuit-breakers to

BS EN 60898 there are unlikely to be circumstances where the armour is not adequate as a protective conductor.

The calculation to determine on an individual basis whether the area of the cable armour is sufficient for the earth fault current can be carried out in either of two ways:

i by plotting the adiabatic characteristics of the sheath or armour on to the overcurrent device characteristics, or

ii by calculation.

Plotting is suitable for preparing standard tables for fuses.

Calculation is suitable for one-off calculations and circuit-breakers.

8.8   Plotting of cable armour adiabatics

Figure 8.2 shows the characteristics of the armour of four-core copper conductor thermosetting insulated cables to BS 5467, plotted over BS 88-2 fuse characteristics. The characteristics are plotted in a similar manner to that described in section 8.6, except that the value for k selected is that from Table 54.4 of BS 7671, e.g. k = 46 for the steel-wire armour of a 90 °C thermosetting insulated cable. The cross-sectional area of the armour is found in Table F.8B in Appendix F.

▼ **Figure 8.2**   Adiabatic characteristics of the armour of four-core copper cables with thermosetting insulation plotted on BS 88-2 fuse characteristics from Figure 3A3(a) of Appendix 3 of BS 7671

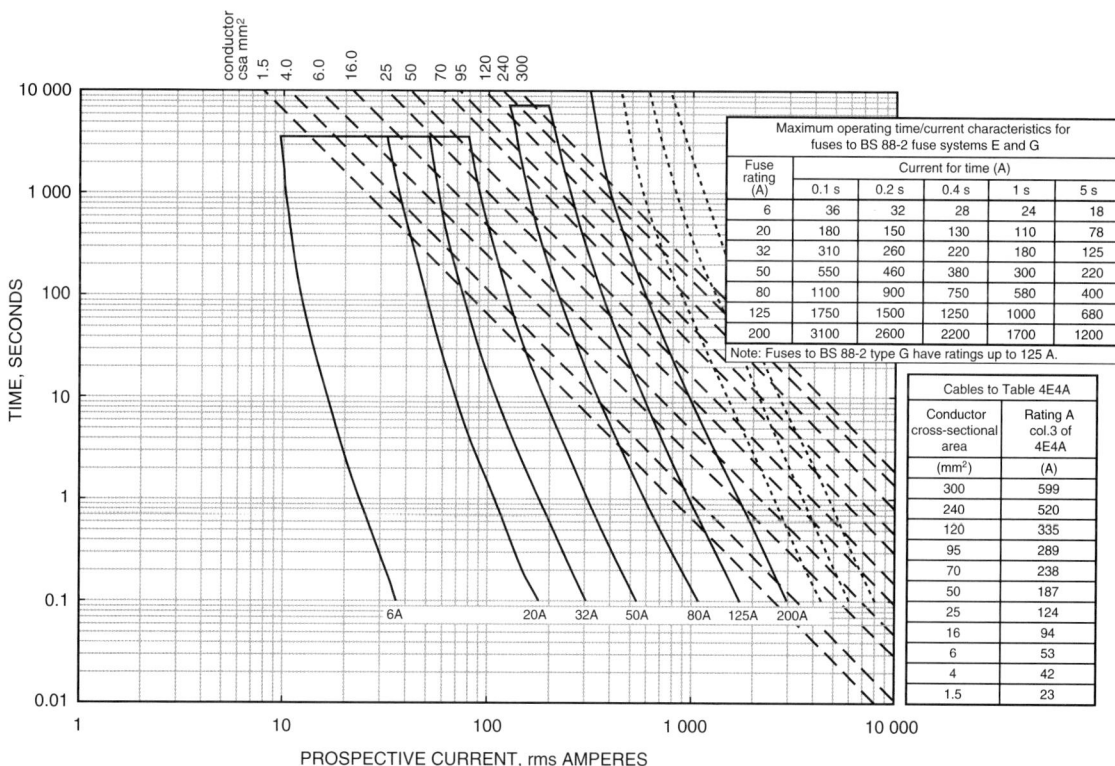

Maximum operating time/current characteristics for fuses to BS 88-2 fuse systems E and G

Fuse rating (A)	Current for time (A)				
	0.1 s	0.2 s	0.4 s	1 s	5 s
6	36	32	28	24	18
20	180	150	130	110	78
32	310	260	220	180	125
50	550	460	380	300	220
80	1100	900	750	580	400
125	1750	1500	1250	1000	680
200	3100	2600	2200	1700	1200

Note: Fuses to BS 88-2 type G have ratings up to 125 A.

Cables to Table 4E4A

Conductor cross-sectional area	Rating A col.3 of 4E4A
(mm²)	(A)
300	599
240	520
120	335
95	289
70	238
50	187
25	124
16	94
6	53
4	42
1.5	23

Example

Consider a four-core 1.5 mm² thermoplastic insulated cable. From Table F.8B the armour cross-sectional area is 18 mm². From Table 54.4 of BS 7671, k is 51 for steel (see Table 8.1 of this guide).

To plot the characteristics we can determine the maximum disconnection time at, say, prospective fault currents of 1000 A, 100 A and 10 A.

From $S = \dfrac{\sqrt{I^2 t}}{k}$, $t = \dfrac{k^2 S^2}{I^2}$.

at 1000 A, $t = \dfrac{18^2\,51^2}{1000^2} = 0.84$ s,   at 100 A, $t = \dfrac{18^2\,51^2}{100^2} = 84$ s,

at 10 A, $t = \dfrac{18^2\,51^2}{10^2} = 8427$ s

The graph in Figure 8.2 indicates that, where the rating of the fuse is less than that of the cable, as is usual for the four-core cables considered, the armour is of sufficient cross-sectional area providing loop impedances are such that disconnection occurs within 5 s. Where fuse ratings exceed the rating of the cable, as may be the case for motor fuses, the check is necessary.

8.9   Calculation of armour capability

If it is wished to determine by calculation whether the armour is of adequate size for the particular installation, the following procedure may be followed:

i   select the overcurrent device,
ii   calculate earth fault impedance and
iii determine either:
    a   the fault current I_{ef} from the earth fault loop impedance, and the time for disconnection t from the characteristics of the overcurrent device, e.g. for BS 88-2 fuses from Appendix 3, Figure 3A3 of BS 7671, or
    b   the maximum energy let-through ($I^2 t$) from the manufacturer's data. This is the usual procedure for circuit-breakers, particularly moulded case type.

Then the equation $S \geq \dfrac{\sqrt{I^2 t}}{k}$ gives the minimum required cross-sectional area of armour.

This can be compared with the values in Table F.8B for cables with thermosetting insulation and Table F.7B for thermoplastic insulated cables.

Example using fuse characteristics

Consider the adequacy of the armour of the submain cable at Board C in the example in Chapter 6 (see Figure 6.4 etc.). The size of the line conductor is 300 mm² aluminium, the overcurrent device a 315 A BS 88-2 fuse.

Z_e is 0.1 ohm, I_{ef} is 2530 A

To use Table 54.7, k_1 and k_2 must be obtained from Tables 43.1 and 54 of BS 7671

k_1 for 300 mm² aluminium cable = 76 from Table 43.1,

k_2 for steel wire armour = 51 (Table 54.4, or from Table 8.1 given earlier)

Minimum cross-sectional area from Table 54.7 is:

$$\frac{k_1}{k_2} \times \frac{S}{2} = \frac{76}{51} \times \frac{300}{2} = 223.5 \text{ mm}^2, \text{ Table F.7B gives 304 mm}^2 \text{ as area of armour, so}$$
is adequate.

Using the adiabatic equation $S = \dfrac{\sqrt{I^2 t}}{k} = \dfrac{\sqrt{2530^2 t}}{51}$

From time/current characteristic for a 315 A fuse with a fault current of 2530 A, t would be less than 5 s; see Table 8.9.

Hence, the required area of armour is given by $S = \dfrac{\sqrt{2530^2 5}}{51} = 111 \text{ mm}^2$

To simplify the selection, a table of minimum area of armour for BS 88-2 fuses is given in Table 8.9. This table assumes disconnection in 5 s. The check is simple. Table 8.9 gives the minimum area of armour for the particular fuse; this can be compared with the values in Table F.7B or F.8B of Appendix F.

▼ **Table 8.9**   Minimum area of steel cable armour when used as a protective conductor in a circuit protected by a BS 88-2 fuse

Fuse rating	Current for 5 s disconnection	Maximum Z_s	Minimum area of armour (mm²)*	
(A)	(A)	(Ω)	Thermosetting	Thermoplastic
6	18	12.8	0.8	0.7
10	32	7.2	1.5	1.3
16	55	4.2	2.7	2.4
20	79	2.9	3.8	3.4
25	100	2.3	4.9	4.3
32	125	1.84	6.1	5.4
40	170	1.35	8.3	7.4
50	220	1.04	10.7	9.6
63	280	0.82	13.6	12
80	400	0.57	19.4	17
100	520	0.44	25.3	24
125	680	0.34	33.5	30
160	820	0.28	44	39
200	1200	0.19	58	52
250	1650	0.14	80	72
315	2200	0.105	107	96
400	2840	0.08	138	124
500	3600	0.064	175	157
630	5100	0.045	248	223
800	7000	0.033	340	307
1000	9500	0.024	462	416
1250	13000	0.018	632	570

* Determined from the adiabatic equation $S = \{\sqrt{(I^2 t)}\}/k$, using k = 46, t = 5 for thermosetting insulation and k = 51, t = 5 for 70 °C thermoplastic, see Table 54.4 of BS 7671 (Table 8.1 in this chapter).

Calculations for Electricians and Designers **101**
© The Institution of Engineering and Technology

▼ **Table 8.10** Maximum BS 88-2 fuse sizes for multicore armoured cables having thermosetting insulation (copper conductors) when the armour is to be used as the protective conductor. Sizes endorsed in Table 4E4A of BS 7671.

BS 5467, BS 6724

Current-carrying capacity

Ambient temperature: 30 °C
Ground ambient temperature: 20 °C
Conductor operating temperature: 90 °C

Conductor cross-sectional area	Reference Method C (clipped direct)		Reference Method E (on a perforated horizontal or vertical cable tray) or Reference 13 (free air)		Reference Method D (direct in ground or in ducting in ground, in or around buildings)		Maximum BS 88 fuse size[3]	
	1 two-core cable, single-phase a.c. or d.c.	1 three- or four-core cable, three-phase a.c.	1 two-core cable	1 three- or four-core cable, three-phase a.c.	1 two-core cable, single-phase a.c. or d.c.	1 three- or four-core cable, three-phase a.c.	Two-core cable	Four-core cable
(mm²)	(A)	(A)	(A)	(A)	(A)	(A)	(A)	(A)
1	**2**	**3**	**4**	**5**	**6**	**7**	**8**	**9**
1.5	27	23	29	25	25	21	63	63
2.5	36	31	39	33	33	28	63	63
4	49	42	52	44	43	36	63	80
6	62	53	66	56	53	44	80	125
10	85	73	90	78	71	58	80	125
16	110	94	115	99	91	75	125	160
25	146	124	152	131	116	96	125	200
35	180	154	188	162	139	115	200	200
50	219	187	288	197	164	135	200	200
70	279	238	291	251	203	167	200	200
95	338	289	354	304	239	197	200	400
120	392	335	410	353	271	223	315	500
150	451	386	472	406	306	251	315	500
185	515	441	539	463	343	281	400	500
240	607	520	636	546	395	324	500	500
300	698	599	732	628	446	365	500	630
400	787	673	847	728	–	–	500	800

Calculations for Electricians and Designers
© The Institution of Engineering and Technology

Notes to Table 8.10:

1 Where a conductor operates at a temperature exceeding 70 °C it shall be ascertained that the equipment connected to the conductor is suitable for the conductor operating temperature (see Regulation 512.1.5).

2 Where cables in this table are connected to equipment or accessories designed to operate at a temperature not exceeding 70 °C, the current ratings given in the equivalent table for 70 °C thermoplastic insulated cables (BS 6004, BS 6346) shall be used (see also Regulation 523.1).

3 The maximum BS 88 fuse size has been calculated assuming the earth fault loop impedance results in disconnection in 5 s.

Table 8.10 shows the maximum BS EN 60269-2 and BS 88-2 fuse sizes for armoured cables with thermosetting insulation to BS 5467 aligned with Table 4E4A of Appendix 4 of BS 7671.

For devices other than fuses to BS 88, in particular moulded case circuit-breakers, the energy let-through (I^2t) can be obtained for the particular device in question from the manufacturer. Knowing k, and the area of the armour, a check of the armour csa is made to determine its suitability for use as the protective conductor, see section 8.5.

8.10 Conduit and trunking

Example

Consider 16 mm^2 thermoplastic insulated cables in 50 x 50 mm steel trunking and calculate if a separate protective conductor is required to supplement the steel enclosure, using Table 54.7. Use k_1 from Table 54.3, as the assumed initial temperature of the line conductor is 70 °C, $k_1 = 115$

k_2 is given by Table 54.5, $k_2 = 47$

$k_1/k_2 = 115/47 = 2.45$

Hence a circuit of 16 mm^2 copper thermoplastic insulated live conductors would require a steel conduit or trunking area of 16 x 2.45 = 39 mm^2, and 50 x 50 mm trunking is equivalent to 135 mm^2 so is more than adequate, see Table 8.11.

The maximum size of line conductor to be included in 50 x 50 mm steel trunking is 135/2.45 = 55 (i.e. 50) mm^2.

Both trunking and conduit almost without exception have sufficient cross-sectional area to meet the requirements of BS 7671. When trunking is used, the manufacturer's data must be obtained to see if there are any limits imposed by the conductivity of connections between lengths of trunking. IET Guidance Note 6 (6.3.6) advises that trunking is not suitable for use as a protective conductor for circuits carrying much more than 100 A.

▼ **Table 8.11** Examples of steel trunking cross-sectional area (BS EN 50085-1)

Nominal trunking size (mm x mm)	Minimum steel cross-sectional area without lid (mm²)
50 x 50	135
75 x 75	243
100 x 50	216
100 x 100	324
150 x 100	378

See Table F.10 in Appendix F for conduit and floor trunking.

543.2.1 Common protective conductors

BS 7671 allows conduit and trunking to be used as a common circuit protective conductor for circuits enclosed within the conduit or trunking.

8.11 Earthing and bonding conductors

542.3 8.11.1 Earthing conductor

543.1 The earthing conductor is a protective conductor which connects the main earthing terminal of the installation to the means of earthing and is sized in the same way as for circuit protective conductors, except that certain minimum cross-sectional areas are specified when the conductor is buried in the ground, see Regulation groups 542.3 and 543.1 and Table 54.1.

The cross-sectional area, subject to the minimum requirements above, of an earthing conductor can be determined using the formula of Regulation 543.1.3:

$$S \geq \frac{\sqrt{I^2 t}}{k}$$

or by using Table 54.7 of BS 7671. The use of Table 54.7 will provide a conservatively sized (oversized) earthing conductor.

Earthing conductors are subject to the minimum sizes of Regulation 543.1.1 as follows:

▶ 2.5 mm² copper equivalent if protected against mechanical damage,
▶ 4.0 mm² copper equivalent otherwise.

Table 8.12 has been prepared on the basis of Table 54.7 for line or neutral conductor sizes of up to 50 mm². For larger sizes the adiabatic equation has been used assuming a fuse size and assuming that the earth fault loop impedance is such as to give 5 s disconnection.

The electricity distributor will normally require a minimum size earthing conductor of 16 mm² for supplies up to 100 A. Electricity distribution systems are excluded from the scope of BS 7671 (Regulation 110.2) and so a disconnection of supply in 5 s for a line to neutral earth fault cannot be assumed.

▼ **Table 8.12** Earthing and main protective bonding conductor sizes (copper equivalent) for TN-S and TN-C-S supplies

Line conductor or neutral conductor of PME supplies (mm²)	4	6	10	16	25	35	50	70	95	120	150	185	240	300
Assumed BS 88 fuse size (A)	Note 6							200	250	315	400	400	500	630
Earthing conductor not buried or buried protected against corrosion and mechanical damage (mm²) [1,2,3]	6	6	10	16	16	16	25	25	35	35	50	50	70	70
Main protective bonding conductor (mm²) [1,4,5]	6	6	6	10	10	10	16	16	25	25	25	25	25	25
Main protective bonding conductor for PME supplies (TN-C-S) (mm²) [4,5]	10	10	10	10	10	10	16	25	25	35	35	50	50	50

Notes:
1 Protective conductors (including earthing and bonding conductors) of 10 mm² cross-sectional area or less shall be copper (Regulation 543.2.4).
2 Electricity distributors may require a minimum size of earthing conductor at the origin of the supply of 16 mm² copper or greater for TN-S and TN-C-S supplies.
3 Buried earthing conductors must be at least:
   25 mm² copper if not protected against mechanical damage or corrosion
   50 mm² steel if not protected against mechanical damage or corrosion
   16 mm² copper if not protected against mechanical damage but protected against corrosion
   16 mm² coated steel if not protected against mechanical damage but protected against corrosion
   (Table 54.1 and Regulation 542.3.1).
4 See Regulation 544.1.1 and Table 54.8.
5 Electricity distributors should be consulted when in doubt.
6 Conductor size determined using Table 54.7 of BS 7671.

544.1 **8.11.2 Main protective bonding conductors**

Adiabatic calculations are not required to determine the cross-sectional area of main protective bonding conductors.

> **544.1.1** Except where PME conditions apply, a main protective bonding conductor shall have a cross-sectional area not less than half the cross-sectional area required for the earthing conductor of the installation and not less than 6 mm². The cross-sectional area need not exceed 25 mm² if the bonding conductor is of copper or a cross-sectional area affording equivalent conductance in other metals.
>
> Except for highway power supplies and street furniture, where PME conditions apply the main protective bonding conductor shall be selected in accordance with the neutral conductor of the supply and Table 54.8.

▼ **Table 54.8 of BS 7671** Minimum cross-sectional area of the main protective bonding conductor in relation to the neutral of the supply.

Note: Local distributor's network conditions may require a larger conductor.

Copper equivalent cross-sectional area of the supply neutral conductor	Minimum copper equivalent* cross-sectional area of the main protective bonding conductor
35 mm² or less	10 mm²
over 35 mm² up to 50 mm²	16 mm²
over 50 mm² up to 95 mm²	25 mm²
over 95 mm² up to 150 mm²	35 mm²
over 150 mm²	50 mm²

* The minimum copper equivalent cross-sectional area is given by a copper bonding conductor of the tabulated cross-sectional area or a bonding conductor of another metal affording equivalent conductance.

8.11.3 Supplementary bonding conductors

544.2 **8.11.3.1 Minimum sizes**

The requirements of Regulation 544.2 are summarized in Table 8.13.

▼ **Table 8.13** Supplementary bonding conductor sizes (mm²)

Size of circuit protective conductor	Minimum cross-sectional area of supplementary bonding					
	Exposed-conductive-part to extraneous-conductive-part		Exposed-conductive-part to exposed-conductive-part		Extraneous-conductive-part to extraneous-conductive-part*	
	Mechanically protected	Not mechanically protected	Mechanically protected	Not mechanically protected	Mechanically protected	Not mechanically protected
	1	2	3	4	5	6
1.0	1.0	4.0	1.0	4.0	2.5	4.0
1.5	1.0	4.0	1.5	4.0	2.5	4.0
2.5	1.5	4.0	2.5	4.0	2.5	4.0
4.0	2.5	4.0	4.0	4.0	2.5	4.0
6.0	4.0	4.0	6.0	6.0	2.5	4.0
10.0	6.0	6.0	10.0	10.0	2.5	4.0
16.0	10.0	10.0	16.0	16.0	2.5	4.0

* If one of the extraneous-conductive-parts is connected to an exposed-conductive-part, the bonding conductor must be no smaller than that required by column 1 or 2.

8.11.3.2 Maximum length of supplementary bonding conductors

415.2

Regulation 415.2.2 requires a further condition to be met:

> **415.2.2** Where doubt exists regarding the effectiveness of supplementary equipotential bonding, it shall be confirmed that the resistance R between simultaneously accessible exposed-conductive-parts and extraneous-conductive-parts fulfils the following condition:
>
> $$R \le \frac{50\,V}{I_a} \text{ in a.c. systems}$$
>
> $$R \le \frac{120\,V}{I_a} \text{ in d.c. systems}$$
>
> where I_a is the operating current in amperes of the protective device:
>
> > for RCDs, $I_{\Delta n}$
>
> > for overcurrent devices, the 5 s operating current.

Part 7 of BS 7671

For Part 7 considerations this condition is almost always complied with, as Regulation 544.2.1 sets a minimum supplementary bonding conductor conductance of that of the smaller or smallest circuit protective conductor.

Consider type C c.bs as the circuit-breakers with high I_a currents; the maximum resistance R and length for various supplementary bonding conductor sizes are calculated in Table 8.14 and, for BS 88-2 fuses, Table 8.15.

Supplementary bonding is normally only applied to final circuits, so larger device ratings have not been considered. Supplementary bonding conductors are unlikely to exceed 10 m in length, so, as can be seen from the data in Tables 8.14 and 8.15, lengths are unlikely to be limited.

▼ **Table 8.14** Type C c.b. – maximum length of supplementary bonding conductors to comply with Regulation 411.4.5

c.b. rating I_n (Note 1) (A)	Current I_a (Note 1) (A)	$R = \dfrac{50}{I_a}$ (Ω)	S_a (Note 2) (mm²)	Conductor resistance, R_2 (Note 3) (mΩ/m)	Maximum length (L) of conductor (area S_a mm²) $L = R \times 1000/R_2$ (m)
6	60	0.83	2.5	7.41	112
10	100	0.5	2.5	7.41	67
16	160	0.312	2.5	7.41	42
20	200	0.25	2.5	7.41	33
25	250	0.20	2.5	7.41	27
32	320	0.156	2.5	7.41	21
40	400	0.125	2.5	7.41	16
50	500	0.10	2.5	7.41	13
63	630	0.079	2.5	7.41	10.7
80	800	0.0625	4.0	4.61	13.55
100	1000	0.05	4.0	4.61	10.8

Notes:
1 From Table to Figure 3A5 of Appendix 3 of BS 7671.
2 From Table 8.4. Minimum copper protective conductor sizes at 6 kA prospective fault current.
3 From Table F.1.

▼ **Table 8.15** BS 88-2 fuse – maximum length of supplementary bonding conductors to comply with Regulation 411.4.5

BS 88 fuse. Rating I_n (A)	Current (0.4 s) I_a (Note 1) (A)	$R = \dfrac{50}{I_a}$ (Ω)	S_a (mm²)	Conductor resistance, R_2 (Note 2) (mΩ/m)	Maximum length (L) of conductor (area S_a mm²) $L = R \times 1000/R_2$ (m)
10	45	1.11	1.5	12.10	91
16	85	0.59	1.5	12.10	48
32	220	0.227	1.5	12.10	18
40	280	0.1768	2.5	7.41	24

Notes:
1 From Tables to Figure 3A3 of Appendix 3 of BS 7671.
2 From Table F.1.

Calculations associated with testing

9

- ■ **General**
- ■ **Continuity**

- ■ **Earth fault loop impedance Z_s**
- ■ **Reduced section protective conductors**

Appx 6 ## 9.1 General

The tests (and measurements) carried out during and on completion of an installation that may require calculations are the measurements for:

▶ continuity and
▶ earth fault loop impedance.

▼ **Figure 9.1** Schedule of test results

Form 4

Form No: ..**1235**...../4

GENERIC SCHEDULE OF TEST RESULTS

DB reference no Consumer Unit	Details of circuits and/or installed equipment vulnerable to damage when testing ELV lights in bathroom	Details of test instruments used (state serial and/or asset numbers)
Location Under-stairs cupboard		Continuity 1012F multi function
Zs at DB (Ω) 0.29		Insulation resistance "
I$_{pf}$ at DB (kA) 0.8 kA		Earth fault loop impedance "
Correct supply polarity confirmed ☑		RCD "
Phase sequence confirmed (where appropriate) N/A		Earth electrode resistance N/A

Tested by:
Name (Capitals) **G THOMPSON**

Signature ....*G.Thompson*.......... Date ..17-Jan-2012...

Circuit details		Overcurrent device			Conductor details			Ring final circuit continuity (Ω)			Continuity (Ω) (R₁ + R₂) or R₂		Insulation Resistance (MΩ)		Polarity	Z$_s$ (Ω)	RCD (ms)		Test button / functionality	Remarks (continue on a separate sheet if necessary)
Circuit number	Circuit Description	BS (EN)	type	rating (A)	breaking capacity (kA)	Reference Method	Live (mm²)	cpc (mm²)	r₁ (line)	r$_n$ (neutral)	r₂ (cpc)	(R₁ + R₂) *	R₂	Live-Live	Live-E			@ I$_{Δn}$	@ 5I$_{Δn}$	Test button / functionality	
1	Ring - sockets downstairs	60898	B	32	6	100	2.5	1.5	0.62	0.62	1.02	0.41	N/A	+299	+299	✔	0.71	28	16	✔	NOTE: The high earth loop impedance
2	Ring - sockets upstairs	60898	B	32	6	100	2.5	1.5	0.62	0.62	1.02	0.41	N/A	+299	+299	✔	0.71	36	21	✔	on the upstairs lighting circuit was
3	Ring - kitchen	60898	B	32	6	100	2.5	1.5	0.22	0.22	0.37	0.15	N/A	+299	+299	✔	0.44	25	18	✔	found to be due to loose terminals
4	Cooker - kitchen	60898	B	32	6	100	6.0	2.5	N/A	N/A	N/A	0.16	N/A	+299	+299	✔	0.46	34	21	✔	at the point of connection of the
5	Lights - downstairs	60898	B	6	6	100	1.5	1.0	N/A	N/A	N/A	2.56	N/A	+299	+299	✔	2.85	29	19	✔	additional lighting circuitry.
6	Lights - upstairs	60898	B	6	6	100	1.5	1.0	N/A	N/A	N/A	8.20	N/A	+299	+299	✔	8.50	33	17	✔	With this additional circuitry
7	Lights - Garage	60898	B	6	6	100	1.5	1.0	N/A	N/A	N/A	1.51	N/A	+299	+299	✔	1.90	31	17	✔	disconnected, a satisfactory
8	Spare																				measurement of 2.86 ohms was
																					obtained. However, the circuit
																					conductors remain damaged and
																					affected cabling must be replaced.

* Where there are no spurs connected to a ring final circuit this value is also the (R₁ + R₂) of the circuit.

Page .4. of .4.

Calculations for Electricians and Designers
© The Institution of Engineering and Technology

109

612.2 9.2 Continuity

Columns 13 and 14 of the Schedule of Test Results (see Figure 9.1) require measurements of $(R_1 + R_2)$ or R_2 to be noted. The measurement of $(R_1 + R_2)$ or R_2 may be simply a continuity test to verify that the line and protective conductors are continuous. $(R_1 + R_2)$ may also be recorded for later use in checking that the earth fault loop impedance is sufficiently low – see section 9.3.

Electricians, having carried out a continuity test and measured $(R_1 + R_2)$ or R_2, need to confirm that the reading is appropriate for the estimated length of cable installed and the ambient temperature. The tables to be used are F.1 and F.2 of Appendix F.

Example

Consider a shower wired in 10 mm² thermoplastic insulated twin cable with 4 mm² protective conductor of length 10 m at an ambient of 10 °C.

The value of $(R_1 + R_2)$ measured is checked against the value $(R''_1 + R''_2)$ in Table F.1 multiplied by the cable length L.

Resistance of cable loop $= (R''_1 + R''_2)$ $(m\Omega/m) \times L(m)$

$$= 6.44 \text{ m}\Omega/\text{m} \times 10 \text{ m} = 64.4 \text{ m}\Omega \text{ or } 0.0644 \text{ }\Omega$$

This is the resistance at 20 °C.

To correct for an ambient temperature of 10 °C Table F.2 is used, and the correction factor given is 0.96.

Reducing the temperature reduces conductor resistance.

Resistance of cable loop corrected to 10 °C = 0.0644 x 0.96 Ω = 0.062 Ω

612.9 9.3 Earth fault loop impedance Z_s

9.3.1 BS 7671 earth fault loop impedance tables

Column 18 of the Schedule of Test Results (Figure 9.1) requires the earth fault loop impedance Z_s to be recorded. The tester then must check that the reading is satisfactory. This will normally be against figures:

▶ provided by the designer, or
▶ from standard circuits, see Chapter 4 of the *Electrician's Guide*, or
▶ standard test values, see Appendix B of the *On-Site Guide*.

This chapter considers how these test figures are calculated.

Tables 41.2 to 41.5 of BS 7671 provide maximum earth fault loop impedances, which must not be exceeded at conductor maximum operating temperature, say 70 °C for thermoplastic insulation. Please see the notes to Tables 41.2 to 41.4; Table 41.3 is given below as a sample.

▼ **Table 41.3 of BS 7671** Maximum earth fault loop impedance (Z_s) for circuit-breakers with U_0 of 230 V, for instantaneous operation giving compliance with the 0.4 s disconnection time of Regulation 411.3.2.2 and 5 s disconnection time of Regulation 411.3.2.3

(for RCBOs see also Regulation 411.4.9)

(a) Type B circuit-breakers to BS EN 60898 and the overcurrent characteristics of RCBOs to BS EN 61009-1

Rating (amperes)	3	6	10	16	20	25	32	40	50	63	80	100	125	I_n
Z_s (ohms)		7.67		2.87		1.84		1.15		0.73		0.46		$46/I_n$
	15.33		4.60		2.30		1.44		0.92		0.57		0.37	

(b) Type C circuit-breakers to BS EN 60898 and the overcurrent characteristics of RCBOs to BS EN 61009-1

Rating (amperes)	6	10	16	20	25	32	40	50	63	80	100	125	I_n
Z_s (ohms)	3.83		1.44		0.92		0.57		0.36		0.23		$23/I_n$
		2.30		1.15		0.72		0.46		0.29		0.18	

(c) Type D circuit-breakers to BS EN 60898 and the overcurrent characteristics of RCBOs to BS EN 61009-1

Rating (amperes)	6	10	16	20	25	32	40	50	63	80	100	125	I_n
Z_s (ohms)	1.92		0.72		0.46		0.29		0.18		0.11		$11.5/I_n$
		1.15		0.57		0.36		0.23		0.14		0.09	

Note: The circuit loop impedances given in the table should not be exceeded when the conductors are at their normal operating temperature. If the conductors are at a different temperature when tested, the reading should be adjusted accordingly. See Appendix 14.

Table 41

9.3.2 Earth fault loop impedance corrections for temperature

When carrying out testing of an installation, the conductors will not be at their operating temperature of say 70 °C for thermoplastic insulated cable. They will be at the site ambient temperature. Consequently, Tables 41.2 to 41.4 in BS 7671 cannot be used for test comparison purposes without adjustment. Resistances at 70 °C are 20% higher than at say 20 °C, so test figures must be lower than the design values (Z_{41}) or those taken from Table 41.2 etc.

The simplest approach is to apply the correction factor C_r also used in Chapter 6. See Table F.3 in Appendix F which shows conductor temperature correction factor C_r, correcting from 20 °C to conductor operating temperature.

For simple conversion to 20 °C, $Z_{test} = Z_{41}/C_r$

For conversion to another ambient (not 20 °C), $Z_{test} = Z_{41}C_{F2}/C_r$

See Table F.2 in Appendix F, which shows the ambient temperature multipliers C_{F2} to be applied to Table F.1 resistances to convert resistances at 20 °C to other ambient temperatures. See also section 10.2.

Example

Consider a 32 A type B circuit-breaker protecting a socket-outlet circuit wired with thermoplastic insulated and sheathed cable.

Table 41.3 of BS 7671 gives a maximum earth fault loop impedance (Z_s, i.e. Z_{41}) of 1.44 ohms.

The correction factor C_r from Table F.3 is 1.2, to correct to 20 °C.

Hence the test loop impedance at 20 °C must not exceed:

$$Z_{test} = Z_{41}/C_r = 1.44/1.2 = 1.2 \ \Omega$$

If it is wished to correct to 10 °C a further correction can be applied and the factors in Table F.2 used, $C_{F2} = 0.96$

$$Z_{test} = Z_{41}C_{F2}/C_r = 1.44 \times 0.96/1.2 = 1.15 \ \Omega$$

To correct more accurately from 70 °C to 10 °C, the formula given in section 10.2 is used:

$$R_{70} = \{1 + 0.004(70 - 10)\}R_{10}$$

$$= 1.24R_{10}$$

In the example, $Z_{test} = Z_{41}/1.24 = 1.44/1.24 = 1.16 \ \Omega$

This is the value given in Table B6 of the *On-Site Guide*, in which a 10 °C ambient is assumed to reflect likely site temperatures.

543   9.4   Reduced section protective conductors

If the protective conductor is of reduced cross-sectional area, that is less than that of the line conductors, a disconnection time less than that necessary for shock protection may be required. This is most likely to occur for fuses in circuits with up to 5 s disconnection times and is unnecessary for circuit-breaker circuits designed so the breaker operates instantaneously.

These calculations are somewhat tedious and whilst they are certainly not beyond the ability of an electrician or a designer they are not something that an electrician would want to carry out on a regular basis when testing. For this reason tables have been prepared for use on site. The fuse tables in Appendix B of the *On-Site Guide* (*OSG*) provide maximum loop impedances not only for each fuse rating, but for each fuse rating and protective conductor combination.

It can be seen from *OSG* Table B2, reproduced below, that for small protective conductor sizes the maximum test loop impedance is further reduced.

For circuit-breakers, the minimum conductor csa including protective conductor is as calculated in section 8.5. Maximum test loop impedances are as calculated in section 9.3.

Chapter 8 on protective conductors provides information as to how these further reductions in loop impedances are calculated.

▼ **Table B2 of OSG** Maximum measured earth fault loop impedance (in ohms) where the overcurrent protective device is a fuse to BS 88-2.2 or BS 88-6. (Testing carried out at an ambient temperature of 10 °C on cables operating at a maximum temperature of 70 °C for various protective conductor sizes.) *

i   0.4 s disconnection (final circuits not exceeding 32A in TN systems)

Protective conductor (mm²)	Fuse rating					
	6 A	10 A	16 A	20 A	25 A	32 A
1.0	6.9	4.1	2.2	1.4	1.2	0.66
1.5	6.9	4.1	2.2	1.4	1.2	0.84
≥ 2.5	6.9	4.1	2.2	1.4	1.2	0.84

ii   5 s disconnection (final circuits exceeding 32 A and distribution circuits in TN systems)

Protective conductor (mm²)	Fuse rating							
	20 A	25 A	32 A	40 A	50 A	63 A	80 A	100 A
1.0	1.7	1.2	0.66	NP	NP	NP	NP	NP
1.5	2.3	1.7	1.1	0.64	NP	NP	NP	NP
2.5	2.3	1.8	1.5	0.93	0.55	0.34	NP	NP
4.0	2.3	1.8	1.5	1.1	0.77	0.50	0.23	NP
6.0	2.3	1.8	1.5	1.1	0.84	0.66	0.36	0.22
10.0	2.3	1.8	1.5	1.1	0.84	0.66	0.46	0.33
16.0	2.3	1.8	1.5	1.1	0.84	0.66	0.46	0.34

*   A value of k of 115 from Table 54.3 of BS 7671 is used. This is suitable for PVC insulated and sheathed cables to Table 4, 7 or 8 of BS 6004 and for thermosetting (LSHF) insulated and sheathed cables to Table 3, 5, 6 or 7 of BS 7211. The k value is based on both the thermoplastic (PVC) and LSHF cables operating at a maximum temperature of 70 °C.

NP Protective conductor, fuse combination NOT PERMITTED.

Impedance of copper and aluminium conductors

10

- ■ **Introduction**
- ■ **Resistance and temperature**
- ■ **Impedance of cables from voltage drop tables**

10.1 Introduction

This chapter discusses in a little more detail than elsewhere in the guide the derivation of the factors for variation of conductor resistance with temperature.

Note to Tables 41.2–41.4

10.2 Conductor resistance and temperature

The resistance of a conductor produces heat in accordance with the equation I^2R. This heat will raise the temperature of the conductor and its insulation and increase the resistance of the conductor. The change of resistance (but not inductance or capacitance) with temperature is significant. The resistance of a copper conductor increases by some 20% if its temperature rises from 20 °C to 70 °C.

Corrections for temperature can be made using the equation:

$$R_t = \{1 + \alpha(t - 20)\}R_{20}$$

where:

R_t = resistance at temperature t
R_{20} = resistance at 20 °C
α = temperature coefficient of resistance.

Values of coefficient of resistance α for common materials at 20 °C are given in Table 10.1.

▼ **Table 10.1** Coefficients of resistance for conductors

Material	Temperature coefficient of resistance α at 20 °C
Annealed copper	0.00393*
Hard drawn copper	0.00381
Aluminium	0.00403*
Lead	0.00400
Steel	0.0045

* An average value of 0.004 is used for both copper and aluminium conductors in temperature correction tables in most publications.

The equation $R_t = \{1 + \alpha(t - 20)\}R_{20}$ becomes $R_t = \{1 + 0.004(t - 20)\}R_{20}$ for copper and aluminium conductors.

Reference to this equation is made below Table F.2 (I2 of the *On-Site Guide*) and is used to prepare Table F.3 (I3 of the *On-Site Guide*).

Example

From $R_t = \{1 + 0.004(t - 20)\}R_{20}$:

The correction factor from 20 °C to 70 °C is $R_{70} = \{1 + 0.004(70 - 20)\}R_{20}$, hence $R_{70} = 1.20\ R_{20}$

The correction factor from 20 °C to 90 °C is $R_{90} = \{1 + 0.004(90 - 20)\}R_{20}$, hence $R_{90} = 1.28\ R_{20}$

The correction factor from 20 °C to 30 °C is $R_{30} = \{1 + 0.004(30 - 20)\}R_{20}$, hence $R_{30} = 1.04\ R_{20}$

These are the values given in Table F.3 and Table I3 of the *On-Site Guide* and are used when calculating conductor resistances at operating temperature from conductor resistances at 20 °C (Table F.1 and Table I1 of the *On-Site Guide*).

10.3 Impedance of cables from voltage drop tables

The voltage drop tables of Appendix 4 of BS 7671 can be used as a source of the resistance and reactance per metre of the cables tabulated.

10.3.1 Single-phase

Tabulated (mV/A/m) voltage drop figures from Appendix 4 of BS 7671 can be used to obtain resistance (at 20 °C) and reactance by:

i correcting from maximum operating temperature to 20 °C, e.g. for thermoplastic insulated cables from 70 °C to 20 °C for the resistance element $(mV/A/m)_r$ only, that is dividing by 1.2, and

ii dividing both the resistance $(mV/A/m)_r$ and inductance $(mV/A/m)_x$ by 2, as the tabulated voltage drop includes voltage drop in the line and the neutral.

▼ **Figure 10.1**
Single-phase circuit

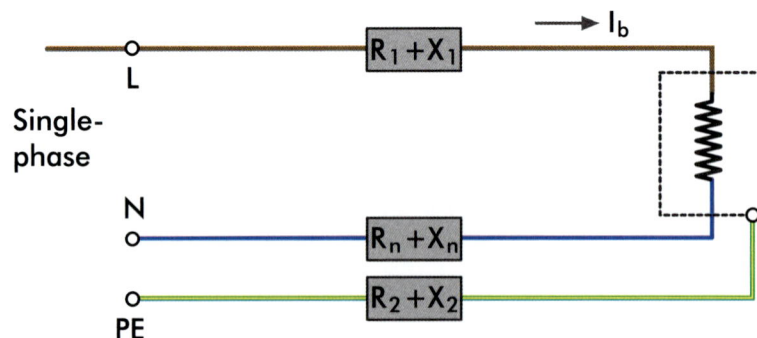

From Figure 10.1 above, voltage drop using $(R''_1 + X''_1)$ is given by

Voltage drop $= L\ I_b\{(C_r R''_1 + X''_1) + (C_r R''_n + X''_n)\}$ in mV

When line and neutral are of the same csa then

Voltage drop $= 2.L\ I_b(C_r\ R''_1 + X''_1)$

where:

L   is the length of the cable (m)
I_b is the load current (A)
$R''_1 + X''_1$ are the resistance and reactance in $m\Omega$ per metre at 20 °C
C_r is the conductor temperature multiplier from Table F.3.

Using the tabulated (mV/A/m), voltage drop $= L\ I_b$(mV/A/m) in mV,
hence $L\ I_b\{$(mV/A/m)$_r$ + (mV/A/m)$_x\}$ in mV is equivalent to $2.L\ I_b(C_r\ R''_1 + X''_1)$
and $\{$(mV/A/m)$_r$ + (mV/A/m)$_x\}$ is equivalent to $2(C_r\ R''_1 + X''_1)$.

Example

From column 3 of Table 4D2B of BS 7671 (see below), for 25 mm² two-core cable:

(mV/A/m)$_r = 1.75$ mV/A/m and (mV/A/m)$_x = 0.170$ mV/A/m

hence $R''_1 = $ (mV/A/m)$/2C_r = 1.75/(2 \times 1.2) = 0.729$ mV/A/m compared with 0.727 Ω/km of Table F.7A and $X''_1 = $ (mV/A/m)$/2) = 0.170/2 = 0.085$ mV/A/m compared with 0.09 Ω/km of Table F.7A.

10.3.2 Three-phase

▼ Figure 10.2

Three- or four-core cable, three-phase a.c. circuit

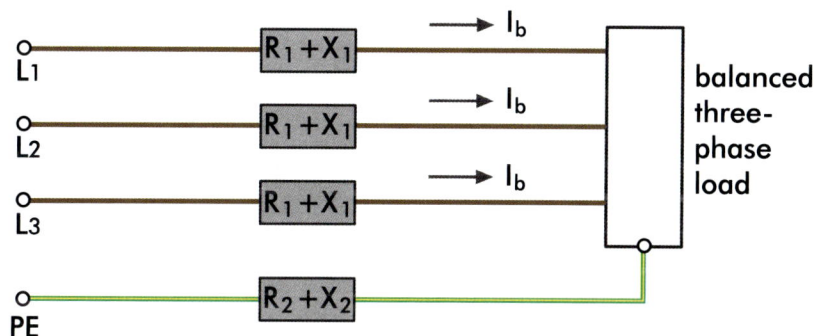

The three- or four-core three-phase a.c. voltage drop figures (mV/A/m) can also be used to obtain resistance (as in Table F.1) and reactance by:

i   correcting from maximum operating temperature to 20 °C, e.g. for thermoplastic insulated cables from 70 °C to 20 °C for the resistance element (mV/A/m)$_r$ only, that is dividing by 1.2, and

ii dividing both the resistance (mV/A/m)$_r$ and reactance (mV/A/m)$_x$ by $\sqrt{3}$, as the tabulated voltage drop is that of the line-to-line voltage.

From Figure 10.2 voltage drop in each line conductor is $L\ I_b(C_r\ R''_1 + X''_1)$

However, the voltage drops in each line conductor are not in phase, so the vector sum across the two lines (line drop) is not twice line but $\sqrt{3}$, hence:

$L\ I_b\{$(mV/A/m)$_r$ + (mV/A/m)$_x\} \equiv \sqrt{3}\ L\ I_b(C_r\ R''_1 + X''_1)$ and
$\{$(mV/A/m)$_r$ + (mV/A/m)$_x\} \equiv \sqrt{3}(C_r\ R''_1 + X''_1)$

▼ Figure 10.3
Vector diagram

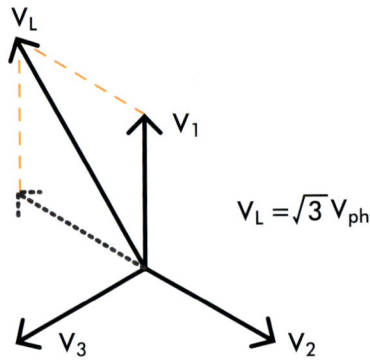

$$V_L = \sqrt{3}\, V_{ph}$$

Note: Line-to-line voltage $V_L = \sqrt{3} V_{ph}$

Example

From Table 4D2B of BS 7671, for 25 mm² three- or four-core cable:

$(mV/A/m)_r = 1.5$ mV/A/m,

hence $R''_1 = (mV/A/m)/(\sqrt{3}C_r) = 1.5/(1.73 \times 1.2) = 0.722$ mV/A/m compared with 0.727 Ω/km of Table F.7A

and $(mV/A/m)_x = 0.145$ mV/A/m,

hence $X''_1 = (mV/A/m)/(\sqrt{3}) = 0.145/(\sqrt{3}) = 0.0837$ mV/A/m compared with 0.09 Ω/km of Table F.1.

▼ Extract from Table 4D2B of BS 7671

VOLTAGE DROP (per ampere per metre):					Conductor operating temperature: 70°C		
Conductor cross-sectional area	Two-core cable, d.c.	Two-core cable, single-phase a.c.			Three-or-four-core cable, three-phase a.c.		
1	2	3			4		
(mm²)	(mV/A/m)	(mV/A/m)			(mV/A/m)		
1	44	44			38		
1.5	29	29			25		
2.5	18	18			15		
4	11	11			9.5		
6	7.3	7.3			6.4		
10	4.4	4.4			3.8		
16	2.8	2.8			2.4		
		r	x	z	r	x	z
25	1.75	1.75	0.170	1.75	1.50	0.145	1.50
35	1.25	1.25	0.165	1.25	1.10	0.145	1.10
50	0.93	0.93	0.165	0.94	0.80	0.140	0.81

Harmonics

11

■ **Introduction**

■ **Cable ratings**

■ **Voltage drop**

■ **Overcurrent protection**

11.1 Introduction

523.1
523.6
Appx 4 sect 5.5

The electronic control and electronic power supplies to much equipment can result in non-linear or non-sinusoidal load current. The basic waveform can be considered to have further waveforms superimposed upon it with frequencies that are multiples of the basic or fundamental waveform. These additional waveforms are called harmonics. Often these harmonics can be disregarded in the design of electrical installations; however, third harmonics and odd multiples of the third harmonic cannot (called triplen harmonics). The triplen harmonic content of discharge lamps can be of the order of 25% with a total harmonic distortion of 30% and the switch-mode power supplies of computers can produce triplen harmonics of the order of 70% with a total harmonic distortion of 77%, and 100% is not unknown.

Normal, that is, fundamental 50 Hz three-phase load currents, if balanced, cancel out in the neutral. This is a natural consequence of the 120-degree time displacement between the phases, see Figure 11.1.

▼ **Figure 11.1**

Phase displacement and third harmonics

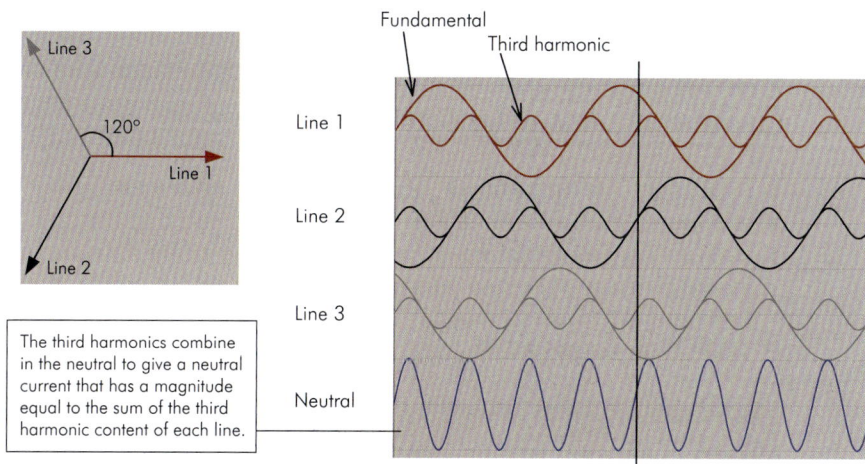

The third harmonics combine in the neutral to give a neutral current that has a magnitude equal to the sum of the third harmonic content of each line.

However, third and other triplen harmonics do not cancel in the neutral but sum so that the neutral current equals the sum of the triplen harmonics of each phase. These harmonic currents can affect:

▶ cable ratings
▶ voltage drop
▶ overcurrent protection
▶ other equipment, e.g. UPS and transformers.

11.2 Cable ratings

523.1
523.6
523.6.100

The tabulated current-carrying capacities in Appendix 4 of BS 7671 are the same for both three- and four-core cables. This is because it is assumed that each line is carrying the rated current and the neutral current is zero or, if the neutral current is not zero as a result of imbalance between the three lines, the current in the neutral will be balanced by a reduction in one or possibly two of the lines. This understanding is not valid if there are harmonics in the line currents, particularly if there are triplen harmonics, that is 3rd, 9th, 15th etc., as these do not cancel out in the neutral but sum. Figure 11.1 shows that the third harmonics in each of the three lines are in phase and will sum in the neutral.

The neutral current I_{bn} is given by $I_{bn} = \dfrac{3h}{100} \times I_{bL}$

where:

I_{bn} is the neutral current from triple harmonics

I_{bL} is the fundamental line current

h is the triple harmonic as a percentage of the fundamental (line) current.

That is, the neutral current is three times the triple harmonic currents in the lines.

The effect of this harmonic neutral current is to derate the cable. If the triple harmonic content exceeds 10%, the neutral conductor should not be of reduced cross-sectional area. If the harmonic content exceeds 15%, the cable must be derated (Tables 11.1 and 11.2).

Table 4Aa

▼ **Table 11.1** Harmonic current rating factors

3rd harmonic content of line current (%)	Neutral current as a percentage of line current (%) [1]	Cable selection [2]	Rating factor [3]
0–15	0–45	Line current	1
>15–33	>45–99	Line current	0.86
>33–45	>99–135	Neutral current	0.86
>45	>135	Neutral current	1

1 Neutral current for 3rd harmonic is three times the line 3rd harmonic.

2 When the neutral current exceeds the line current, selection is based on the neutral current.

3 Harmonic currents reduce the rating of a cable. Where the neutral current exceeds 135% of the line current and selection is based on the neutral current, no rating factor is applied.

▼ **Table 11.2** Calculation of current-carrying capacity

Load current, I_{bL} (A)	Triple harmonic content, h (%)	Neutral current, I_{bn} (A) *	Rating selection basis	Current-carrying capacity required, I_z (A)
100	0–15	0–45	I_{bL}	100
100	>15–33	>45–99	$I_{bL}/0.86$	116
100	>33–45	>99–135	$I_{bn}/0.86$	>116–157
100	>45	>135	I_{bn}	>135

* $I_{bn} = 3h/100\ I_{bL}$ for triple harmonics

Example 1

Consider a load of 200 A with a harmonic content of 20% to be supplied by a four-core cable on a cable tray.

Table 11.2 advises a cable of rating 200/0.86 be selected, that is 233 A.

Table 4D4A of BS 7671 requires a 95 mm² cable. Without harmonics a 70 mm² cable would be suitable.

Example 2

Consider a load of 200 A with a harmonic content of 40%.

The neutral current $I_{bn} = \dfrac{3 \times 40}{100} \times 200 = 240$ A

Table 11.1 advises a cable rating of (neutral current)/0.86, therefore cable rating = 240/0.86 = 279 A.

Table 4D4A of BS 7671 requires a 120 mm² cable.

11.3 Voltage drop

Triple harmonics have a compound effect on voltage drop. As well as producing increased voltage drop due to the current in the neutral, the voltage drop is also increased because the triple harmonics increase the effective inductance of the cable. Inductive reactance (2πfL) is proportional to frequency: the higher the frequency the higher the inductive reactance.

The general equation for voltage drop from Appendix 4 of BS 7671 is as below:

$$\text{Voltage drop} = \frac{LI_b}{1000}\{\cos\varnothing\,(mV/A/m)_r + \sin\varnothing\,(mV/A/m)_x\}$$

where:

L   is the length of the cable
I_b is the load
cos Ø   is the power factor
$(mV/A/m)_r$ and $(mV/A/m)_x$ are voltage drop values (in mV per amp per metre) given in Appendix 4 for the appropriate cable.

For cable sizes up to 16 mm² the simplified formula can be used as follows:

$$\text{Voltage drop} = \frac{LI_b}{1000}\{\cos\varnothing\,(mV/A/m)_r\}$$

These formulae when used for three-phase circuits assume a balanced load; that is, a negligible neutral current. For a load with a high third harmonic content the neutral current is not zero and the inductive reactance of the cable increases. The revised voltage drop formulae are given below.

Cable sizes larger than 16 mm²

$$\text{Voltage drop} = \frac{LI_b}{1000}\{\cos\emptyset\ (\text{mV/A/m})_r\left(1+\frac{3h}{100}\right) + \sin\emptyset\ (\text{mV/A/m})_x\left(1+\frac{11h}{100}\right)\}$$

(**Note:** Readers proving this equation for themselves will note the neutral current is three times the line third harmonic, the neutral inductive voltage drop is trebled, and not to be forgotten the line voltage drop is also slightly increased as its inductive reactance to the third harmonic element is trebled.)

Cable sizes 16 mm² and smaller

$$\text{Voltage drop} = \frac{LI_b}{1000}\{\cos\emptyset\ (\text{mV/A/m})_r\left(1+\frac{3h}{100}\right)\}$$

Examples with cable size over 16 mm²

Consider again the load of 200 A with a third harmonic content of 20% to be supplied by a four-core cable on a cable tray.

Table 11.2 advises a cable of rating 200/0.86 be selected, that is 233 A.

Table 4D4A of BS 7671 says a 95 mm² cable should be selected (and not 70 mm² if the third harmonic was neglected). Assume 50 m length and a power factor of 0.8.

$$\text{The voltage drop} = \frac{50\times200}{1000}\{0.8\times0.41\left(1+\frac{3\times20}{100}\right) + 0.6\times0.135\left(1+\frac{11\times20}{100}\right)\}$$

The voltage drop = 7.8 V as compared with 4.1 V if third harmonics are neglected.

The effect is most pronounced for large single-core cables, as they have a relatively high inductance:

Consider four 630 mm² single-core copper cables (BS 7671 Table 4E1) of length 50 m, laid flat and touching, supplying a load of 1000 A with a third harmonic content of 20% at a power factor of 0.9.

$$\text{The voltage drop} = \frac{50\times1000}{1000}\{0.9\times0.071\left(1+\frac{3\times20}{100}\right) + 0.44\times0.160\left(1+\frac{11\times20}{100}\right)\}$$

The voltage drop = 16.4 V as compared with 6.7 V if third harmonics are neglected.

11.4 Overcurrent protection

High harmonic currents in the load do not affect fault current calculations as fault currents are generally determined by circuit characteristics and not load characteristics, on the presumption that the fault currents are significantly higher than load currents. If this is not the case then allowances will need to be made.

However, for overload protection this is not so.

The usual formula, $I_z \geq I_n \geq I_b$, is valid for triple harmonic content up to 15%. For greater harmonic content, selection can be made as follows, where the device rating I_n installed in the line conductors has been selected on the basis of the line current I_b:

for triple harmonic content	0–15%	$I_z \geq I_n$
for triple harmonic content	>15–33%	$I_z \geq I_n/0.86$
for triple harmonic content	>33–45%	$I_z = \dfrac{3h}{86} I_n$
for triple harmonic content	above 45%	$I_z = \dfrac{3h}{100} I_n$

and $I_n > I_b$

where:

I_n is the rating of the overcurrent device in the line conductors
h is the triple harmonic as a percentage of the fundamental (line) current
I_z is the current-carrying capacity of the cable under particular installation conditions.

Overcurrent protection may be provided by devices in the line conductor; however, it may be appropriate to fit overcurrent detection in the neutral which must disconnect the line conductors, but not necessarily the neutral, see Regulation 431.2.3.

For PEN conductors in TN-C or TN-C-S systems the PEN conductor must not be switched (Regulation 537.1.2). It may be appropriate to fit an overcurrent device in the neutral. However, this must disconnect the line conductors and will not necessarily provide overload protection unless carefully selected with a knowledge of the harmonic current.

With a triple harmonic content exceeding 33% of the fundamental, neutral currents can exceed the line currents. There are then certain attractions in fitting the overcurrent detection in the neutral conductor; however, this overcurrent detection must disconnect the line conductors and care must be taken in adopting this approach as there is a presumption that the harmonic content will remain constant over the life of the installation. It is perhaps preferable to degrade the overcurrent protection in the line conductors accordingly; this is more of a fail-safe approach.

Example

Consider a load of 100 A with a harmonic content of 50%.

This load can be protected by 100 A devices fitted in the line conductors. The neutral current I_{bn} is given by:

$$I_{bn} = \frac{3h}{100} I_{bL} = \frac{3 \times 50}{100} \times 100 = 150 \text{ A}$$

The cable rating I_z must therefore be at least 150 A.

Protection against voltage disturbances   **12**

- ■ **Introduction**
- ■ **The overvoltages**
- ■ **Power frequency fault voltage**
- ■ **Power frequency stress voltages**
- ■ **Earthing of 11 kV substations**

12.1 Introduction

This chapter considers the requirements for the protection of low voltage installations against temporary overvoltages. The requirements are found in Chapter 44 of BS 7671 and also in BS EN 61936-1: *Power installations exceeding 1 kV*, and BS EN 50522: *Earthing of power stations exceeding 1kV*, particularly the UK annexes. Designers need to make reference to these Standards.

442.1 ## 12.2 The overvoltages

BS 7671:2008 considers four situations which generally cause the most severe temporary overvoltages:

- ▶ faults between the high voltage system and Earth
- ▶ loss of the supply neutral in a low voltage system
- ▶ short-circuits in the low voltage installation
- ▶ accidental earthing of a low voltage IT system

Designers need to consult with the electricity distributor to obtain the following information with respect to the high voltage system:

- ▶ the quality (reliability) of the neutral earthing
- ▶ the maximum level of earth fault current
- ▶ the resistance of the earthing arrangement.

However, the UK annexes to BS EN 50522 reduce the tolerable touch voltages, see Figure 12.1 below.

12.3 Power frequency fault voltage

The magnitude and duration of the fault voltage U_f which appears in the low voltage (LV) installation between exposed-conductive-parts and Earth shall not exceed the values given by the curve in Figure 44.2 of BS 7671 for the duration of the fault.

If the PEN conductor of the low voltage system is connected to Earth at more than one point it is permitted to double the value of U_f given in Figure 12.1.

▼ **Figure 12.1**
(Figure 44.2 of BS 7671) Tolerable fault voltage due to an earth fault in the HV system

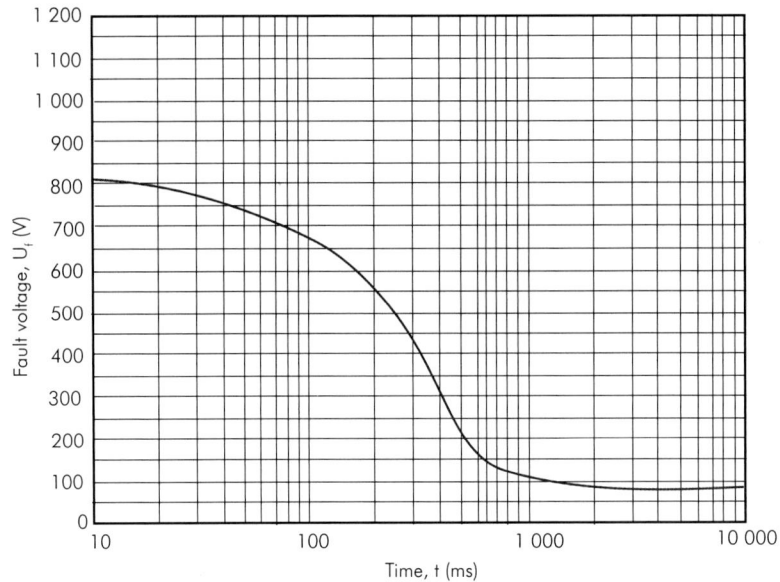

This chapter only considers TN-C (including TN-C-S) and TN-S systems. Whilst TT systems are used in the UK they are not generally applicable to the industrial installations or large commercial installations which designers will normally encounter. Should the designer be considering TT or IT systems then more detailed reference will need to be made to the IEC Standards.

The requirements with respect to this chapter for TN-C-S systems are the same as for TN-C systems.

U_f, the power frequency fault voltage for TN-C, TN-C-S and TN-S, is given by:

$U_f = R_E \times I_E$ for common HV and neutral earths as Figures 12.2 and 12.4
$U_f = 0$ for separated HV equipment and LV neutral earths as per Figures 12.3 and 12.5.

Consideration needs to be given to the use of common HV equipment/LV neutral earths, as this is most common. It can be impractical to separate HV equipment and LV neutral earths in normal underground distribution systems.

A requirement of the Electricity Supply Regulations, replaced by the Electricity Safety, Quality and Continuity Regulations 2002, was that where there was a combined HV equipment and neutral earth then the resistance to earth should not exceed 1 ohm. The Electricity Safety, Quality and Continuity Regulations, however, have a non-specific requirement.

This condition that U_f meet the requirements of national annex NA1 of BS EN 50522 is based on the simple worst case where the low voltage system neutral conductor is earthed only at the transformer substation earthing arrangement. Where, in compliance with the relevant requirements of BS EN 61936-1, either the PE or PEN conductor is earthed at several points or the earthing is part of a global earthing system (see 2.7.19 of IEC 61936-1) the tolerable U_f is twice that given in Figure 12.1.

For a TN-S installation carried out in an industrial site, the designer will normally multiply-earth the distribution system protective conductor. Where each submain enters a separate building an additional earth will need to be installed to provide for 'global earthing'.

Notes to Figures 12.2 to 12.5:

I_E is that part of the earth fault current in the high voltage system that flows through the earthing arrangement of the transformer substation

R_E is the resistance of the earthing arrangement of the transformer substation

R_B is the resistance of the earthing arrangement of the low voltage system neutral, for low voltage systems in which the earthing arrangement of the transformer substation and of the low voltage system neutral are electrically independent

U_0 is the line-to-neutral voltage of the low voltage system

U_f is the voltage which appears in the low voltage system between exposed-conductive-parts and Earth for the duration of the fault

U_1 is the power frequency stress voltage in the low voltage equipment of the transformer substation

U_2 is the power frequency stress voltage in the low voltage equipment of the low voltage installation.

▼ **Figure 12.2**
TN-C system common HV equipment and LV neutral earth, with HV fault to earth

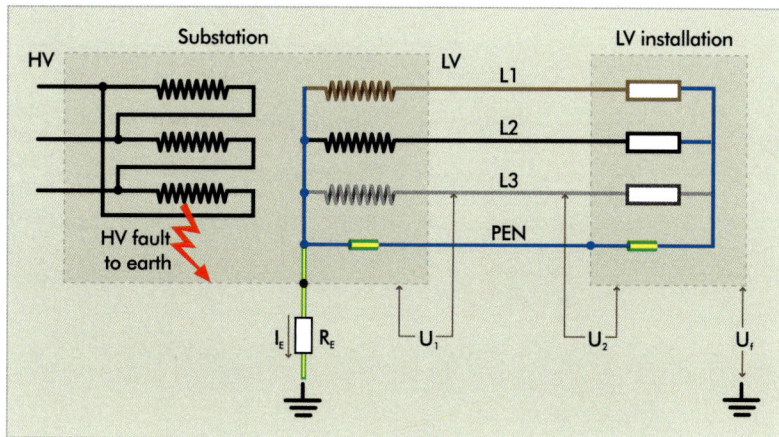

$$U_1 = U_0 \qquad U_2 = U_1 = U_0 \qquad U_f = R_E \times I_E$$

▼ **Figure 12.3**
TN-C system separated HV equipment and LV neutral earth, with HV fault to earth

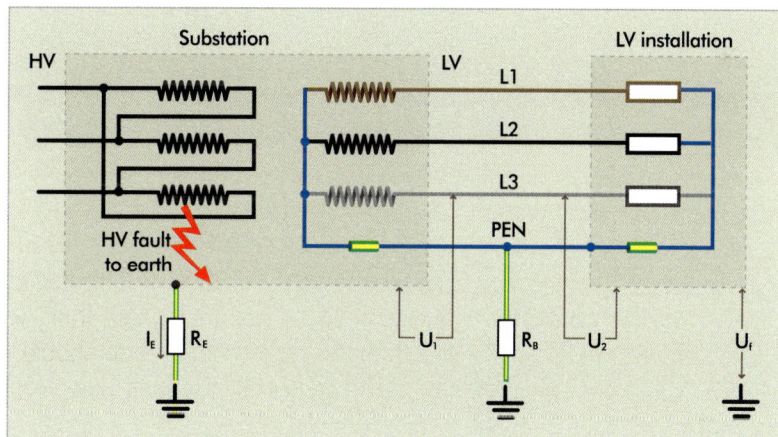

$$U_1 = R_E \times I_E + U_0 \qquad U_2 = U_0 \qquad U_f = 0$$

▼ Figure 12.4

TN-S system common HV equipment and LV neutral earth, with HV fault to earth

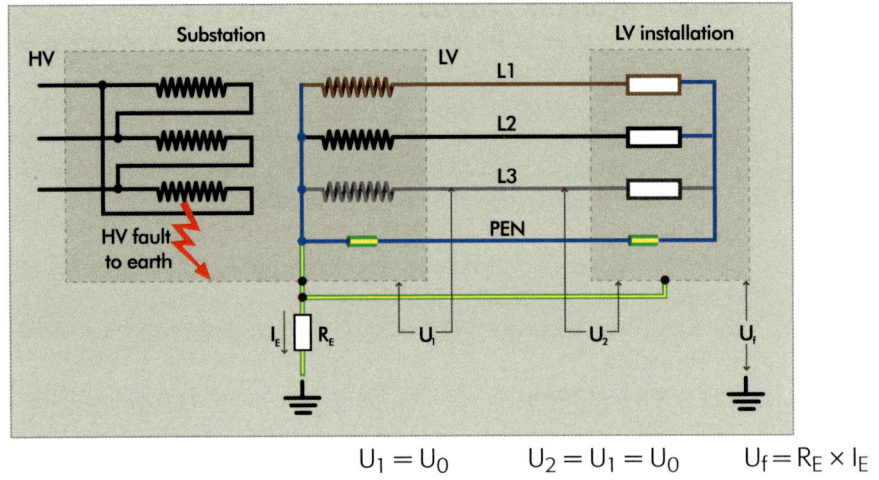

$$U_1 = U_0 \qquad U_2 = U_1 = U_0 \qquad U_f = R_E \times I_E$$

▼ Figure 12.5

TN-S system separated HV equipment and LV neutral earth, with HV fault to earth

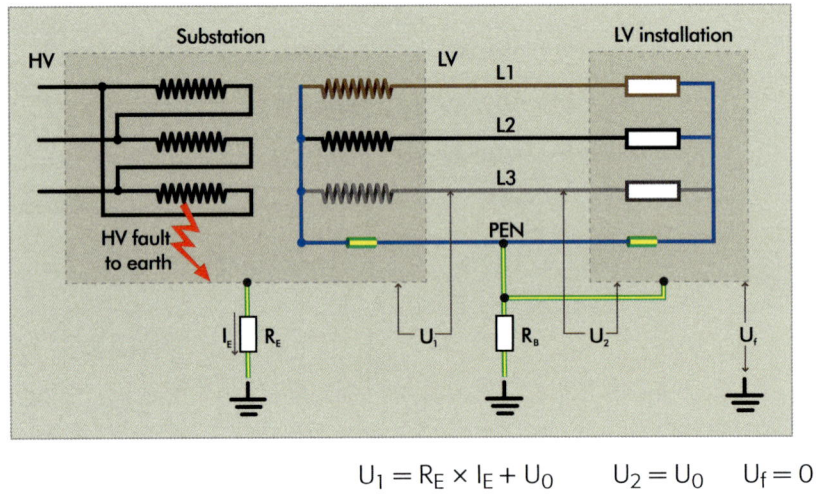

$$U_1 = R_E \times I_E + U_0 \qquad U_2 = U_0 \qquad U_f = 0$$

Calculations for Electricians and Designers
© The Institution of Engineering and Technology

12.4 Power frequency stress voltages

The magnitude and duration of the power frequency stress voltages given by U_1 and U_2 in the figures are required not to exceed the values of Table 12.1.

▼ **Table 12.1** Permissible stress voltages U_1 and U_2 (Table 44.2 of BS 7671)

Permissible power frequency stress voltage on equipment in low voltage installations, U	Duration of earth fault in the high voltage system
$U_0 + 250$ V	> 5 s
$U_0 + 1200$ V	≤ 5 s
In IT systems U_0 shall be replaced by the line-to-line voltage.	

Notes:

1 The requirements in respect of the power frequency stress voltage for the low voltage equipment of the transformer substation are given in section 12.3.

2 The first line of the table relates to high voltage systems having long disconnection times, for example, isolated neutral and resonant earthed high voltage systems. The second line relates to high voltage systems having short disconnection times, for example low-impedance earthed high voltage systems. Both lines together are relevant design criteria for insulation of low voltage equipment with regard to temporary power frequency overvoltage (see 3.3.3.2.2 of IEC 60664.1).

3 In a system whose neutral is connected to the earthing arrangement of the transformer substation, such temporary power frequency overvoltage is also to be expected across insulation which is not in an earthed enclosure when the equipment is outside a building.

In TN systems with a common HV/LV earth, U_1 and U_2 will not exceed U_0, so no special precautions need be taken. However, when the neutral conductor is earthed via an earth arrangement electrically independent of the earthing arrangement of the transformer substation (see Figures 12.3 and 12.5 for TN systems), the insulation level of the low voltage equipment of the transformer substation shall be compatible with the power frequency stress voltage ($R_E \times I_E + U_0$).

Note: The insulation level of the LV equipment of the transformer substation may be higher than the value given in Table 12.1.

12.5 Earthing of 11 kV substations

Requirements for high voltage installations are given in BS EN 61936-1 *Power installations exceeding 1 kV AC – Part 1: Common rules*. More detailed requirements for earthing high voltage installations are given in BS EN 50522 *Earthing of power installations exceeding 1 kV*. A number of national annexes deal with the particular requirements for the United Kingdom.

The requirements of BS EN 50522 are probably more onerous than those of BS 7671 and must be complied with. The Health and Safety Executive have supported the requirements as a means of complying with the Electricity Safety, Quality and Continuity Regulations.

12

Busbar trunking   13

- ■ **Symbols**
- ■ **Voltage drop**

- ■ **Fault currents**

13.1 Symbols

The calculations for busbar trunking are considered separately to cables in conduit and trunking as manufacturers' data is usually presented differently. The symbols used by manufacturers for busbar trunking are often those of CENELEC report R064-003 (see Appendix E). The symbols used in this guide are those of BS 7671 and IET guides, but in this chapter the CENELEC symbols are also listed.

13.2 Voltage drop

Values for trunking voltage drop (mV/A/m) are often given for specific load power factors, and if this is known the most appropriate from Table 13.1 is selected.

▼ **Table 13.1**   Typical* busbar trunking voltage drop (in mV/A/m) for three-phase 50 Hz current, with load distributed along the run. If the load is concentrated at the end of the run, voltage drops are double the values indicated.

Load power factor Ø	Type of trunking:	KVC-20	KVC-31	KVC-40	KVC-63	KVC-80
	Rating (A):	200	315	400	630	800
Ø = 0.7		0.410	0.168	0.154	0.072	0.0617
Ø = 0.8		0.467	0.190	0.175	0.086	0.0696
Ø = 0.9		0.523	0.211	0.194	0.089	0.077
Ø = 1		0.577	0.231	0.212	0.094	0.084

*   Canalis KVC (200 to 800 A).

Example 1

Consider a 300 A three-phase load at an assumed 0.8 power factor evenly distributed over a 20 m length of 315 A trunking.

Voltage drop = 300 A x 20 m x 0.190 mV/A/m = 1140 mV = 1.14 V

Example 2

Consider a 300 A three-phase load at an assumed 0.8 power factor concentrated at the end of the run of a 20 m length of 315 A trunking.

Voltage drop = 300 A x 20 m x 2 x 0.190 mV/A/m = 2280 mV = 2.28 V

13.3 Fault currents

In Table 13.2, resistance and reactance are given for a particular medium power busbar system.

As well as resistance at 20 °C (R_{b0ph}), the table also gives resistance at full-load temperature (R_{b1ph}) and at the average of the temperature of the busbar at full load and that reached under short-circuit conditions (R_{b2ph}).

▼ **Table 13.2** Medium power busbar trunking

CENELEC R064-003 symbols		KVC-20	KVC-31	KVC-40	KVC-63	KVC-80
Rated current I_{th} (A)		200	315	400	630	800
Rated peak current (kA)		26	48	52	76	82
Permissible rated short time (kA r.m.s.)		7.2	18	19	31	35
Live conductors	BS 7671 symbol					
R_{b0ph} Average resistance per conductor, cold state (mΩ/m) (ambient temperature 20 °C)	R_1	0.539	0.216	0.198	0.086	0.078
R_{b1ph} Average resistance per conductor at full load I_{th} (mΩ/m) (ambient temperature 35 °C)	$C_r R_1$	0.666	0.266	0.245	0.106	0.0967
X_{ph} Average reactance per conductor (mΩ/m)	X_1	0.01	0.01	0.01	0.01	0.005
Z_{ph} Average impedance per conductor (mΩ/m)	Mod ($C_r R_1 + X_1$) or Z_1 at maximum load	0.666	0.266	0.245	0.106	0.0967
$R_{b1ph\ ph}$ or $_{phN}$ Average resistance of fault loops between live conductors (mΩ/m)	$2 C_r R_1$	1.332	0.532	0.49	0.212	0.193
$R_{b2ph\ ph}$ or $_{phN}$ Average resistance of fault loops between live conductors (mΩ/m)	$2 \times 1.48 R_1$	1.598	0.638	0.588	0.254	0.232
$R_{b1ph\ pe}$ Average resistance of fault loops between live conductors and PE (mΩ/m)	$C_r(R_1 + R_2)$	1.33	0.93	0.784	0.645	0.425
$X_{b1ph\ ph}$ or $_{phN}$ Average reactance of fault loops between live conductors (mΩ/m)	$2X_1$	0.02	0.02	0.02	0.02	0.01
$X_{b1ph\ pe}$ Average reactance of fault loops between live conductors and PE (mΩ/m)	$X_1 + X_2$	0.02	0.02	0.02	0.02	0.01

Notes:

1 For 60 to 400 Hz a.c. or for d.c. please consult your regional customer centre.

2 Rated currents I_{th} are given for an average ambient temperature of 35 °C and for a casing temperature rise of no more than 55 °C according to the test conditions of standard IEC 439.2.

Since the drafting of the CENELEC document, BS 7671 has been amended to require earth fault loop impedances at full-load temperature only (see notes to Tables 41.2 to

41.4), and not the average of the temperature of the conductor at full load and that reached under short-circuit conditions. It was considered that such detailed corrections were unnecessary and probably less accurate.

▼ **Figure 13.1**
Example busbar installation

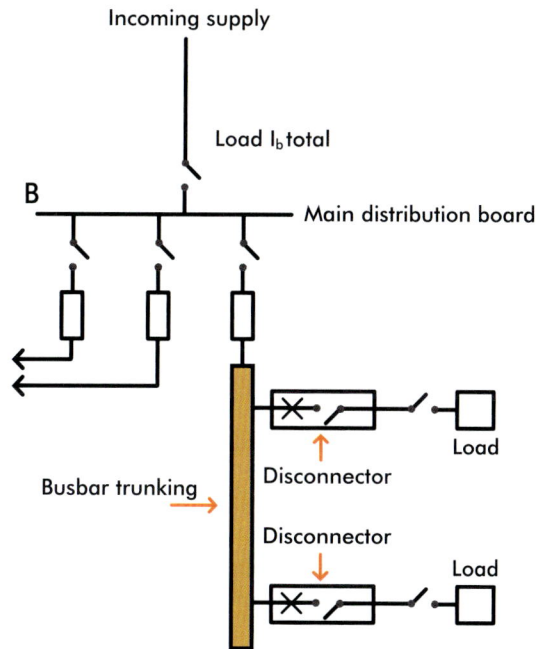

Example

50 metres of 315 A busbar is to be supplied from a distribution board, see Figure 13.1 above. The board is assumed to be board B of the distribution system shown in Figure 6.4 of Chapter 6.

To calculate the three-phase fault current at the origin of the busbar where the busbar impedance has no effect:

$$I_{pf} = \frac{U_{oc}}{Z_x + Z_D} \cong \frac{C_{max} U_0}{Z_x + Z_D}$$

1. Three-phase to earth fault current at B

	Impedance	
	r (Ω)	**x (Ω)**
500 kVA transformer (Table F.5)	0.0051	0.0171
25 metres of 600 mm² single-core aluminium armoured (Table F.9) line impedance 25 x (0.0515 r + 0.09 x)/1000 Ω	0.0013	0.0023
Temperature correction (Note 1)	–	–
Total line impedance at B (Z_{pf})	0.0064	0.0194

Note 1: Temperature correction is not applied for three-phase to earth faults where the worst condition is a 'cold' installation.

$$Z_{pf} = \sqrt{r^2 + x^2} = \sqrt{0.0064^2 + 0.0194^2} = 0.0204 \ \Omega$$

$$I_{pf} = 1.1 \times 230/0.0204 = 12\ 402 \ A$$

2. Three-phase to earth fault current at end of busbar, 20 °C temperatures

To calculate the three-phase fault current at end of the busbar:

$$I_{pf} = \frac{C_{max} U_0}{Z_x + Z_D + R_1 + X_1}$$

	Impedance	
	r (Ω)	**x (Ω)**
Line impedance at B from 1 above	0.0064	0.0194
50 metres of 315 A busbar line impedance Table 13.2 $50 \times (0.216\,r + 0.01\,x)/1000\ \Omega$	0.0108	0.0005
Temperature correction (Note 1)	–	–
Total line impedance at end of busbar (Z_{pf})	0.0172	0.0199

Note 1: Temperature correction is not applied for three-phase to earth faults where the worst condition is a 'cold' installation.

$$Z_{pf} = \sqrt{r^2 + x^2} = \sqrt{0.0172^2 + 0.0199^2} = 0.0263\ \Omega$$

$$I_{pf} = 1.1 \times 230/0.0263 = 9\,620\ A$$

3. Earth fault loop impedance at end of busbar

	Impedance	
	r (Ω)	**x (Ω)**
Transformer	0.0051	0.0171
25 m of 600 mm² aluminium singles (Table F.9)		
(a) line impedance at 20 °C	0.0013	0.0023
(b) neutral/earth impedance	0.0013	0.0023
(c) correction to 70 °C (a + b) x 0.20 (Note 2)	0.0005	–
50 metres of 315 A busbar fault loop impedance Table 13.2 $50 \times (0.93\,r + 0.02\,x)/1000\ \Omega$	0.0465	0.0010
Temperature correction – included in $Rb_{1ph\ pe}$	–	–
Total earth fault loop impedance at end of busbar	0.0547	0.0227

Note 2: Correction factor from 20 °C to 70 °C for an aluminium conductor is $(70 - 20) \times 0.004 = 0.20$ (see Table F.4). The correction factor is only applied to the resistive component of the impedance.

$$Z_{ef} = \sqrt{0.0547^2 + 0.0227^2} = 0.059\ \Omega$$

$$I_{ef} = 1.1 \times 230/0.059 = 4\,288\ A$$

From Table F.19, this fault current is sufficient to operate a 315 A BS 88-2 fuse at B within 5 s. Manufacturers' moulded case circuit-breaker data would be consulted for other devices.

Symbols

A

The symbols used in this guide are summarized below. Where the symbol is used in BS 7671 it has the same description. There are symbols included in the publication that are not used in BS 7671. The major use in this guide is referenced and symbols not found in BS 7671 are asterisked *.

Symbol		Symbol description	Unit	Major use
α	*	Temperature coefficient of resistance		10.2
C		Rating factor – general		4.0
c	*	Voltage factor, ratio between the equivalent voltage source and the nominal system voltage U_0 divided by $\sqrt{3}$ (see Table A.1)		
C_a		Current-carrying capacity rating factor for ambient temperature		4.3
C_c		Current-carrying capacity rating factor for buried circuits		4.3
C_d		Current-carrying capacity rating factor for depth of buried circuits		4.3
C_f		Current-carrying capacity rating factor where $I_2 > 1.45\ I_n$		4.3
C_g		Current-carrying capacity rating factor for grouping		4.3
C_i		Current-carrying capacity rating factor for conductors embedded in thermal insulation		4.3
C_r	*	Rating factor to correct from a resistance at 20 °C to conductor operating temperature		2.6
C_t		Voltage drop factor for operating temperature of conductor		2.5
C_{F2}	*	Conductor resistance correction factor for ambient temperature		9.3.2
cos Ø		Power factor (sinusoidal)		3.3
D_e		Overall cable diameter		D.4
f		Frequency in cycles per second	Hz	2.2.1
gG		Class 'gG' utilization category of fuses to BS 88 – general use		F.19
gM		Class 'gM' utilization category of fuses to BS 88 motor circuit applications		F.19
h		The triple harmonic as a percentage of the fundamental (line) current		11.2

Symbol		Symbol description	Unit	Major use
I		Current (general term)	A	3.3.1
I_a		Current causing automatic operation of protective device within the time stated	A	7.2.2
I_b		Design current of circuit	A	2.4.1
I_{bn}		Neutral current from triple harmonics	A	11.2
I_{bL}	*	Fundamental line current	A	11.2
I_{ef}	*	Prospective earth fault current	A	6.3.2
I_n		Rated (nominal) current or current setting of protective device	A	2.4.1
$I_{\Delta n}$		Rated residual operating current of the protective device in amperes	A	8.11.3.2
I_{pf}		Prospective fault current	A	2.2.1
I_t		Tabulated current-carrying capacity of a cable	A	2.4.1
I_z		Current-carrying capacity of a cable for continuous service, under the particular installation conditions concerned	A	2.4.1
I^2t		Energy let-through value of device	A^2s	8.5
I_2		Current causing effective operation of the overload protective device	A	4.2.3
k		Material factor taken from Tables 43.1 and 54.2 to 54.6 of BS 7671	$As^{1/2}mm^{-2}$	8.1
(mV/A/m)		Voltage drop per ampere per metre	$mVA^{-1}m^{-1}$	2.5
$(mV/A/m)_r$		Resistive voltage drop per ampere per metre	$mVA^{-1}m^{-1}$	5.4
$(mV/A/m)_x$		Reactive voltage drop per ampere per metre	$mVA^{-1}m^{-1}$	5.4
$(mV/A/m)_z$		Impedance voltage drop per ampere per metre	$mVA^{-1}m^{-1}$	5.4
L	*	Cable length	m	2.6
L_a	*	Maximum cable length if the limitation of the adiabatic equation is to be met	m	2.8
L_s	*	Maximum length of cable in the circuit if the disconnection requirements in the event of a fault are to be met	m	2.6
L_{vd}	*	Cable length if the voltage drop limit is not to be exceeded	m	2.5
P	*	Power	VA	3.3
R		Resistance of supplementary bonding conductor	Ω	8.11.3.2
R_1		Resistance of line conductor of a distribution or final circuit	Ω	2.6
R''_1	*	Resistance per metre of line conductor of a distribution or final circuit	Ω/m	2.6
R_2		Resistance of circuit protective conductor (cpc) of a distribution or final circuit	Ω	2.6

Symbol		Symbol description	Unit	Major use
R''_2	*	Resistance of circuit protective conductor (cpc) per metre of a distribution or final circuit	Ω/m	2.6
R_{20}	*	Resistance of a conductor at 20 °C	Ω	10.2
S		Nominal cross-sectional area of the conductor	mm^2	8.1
t		Duration of the fault in seconds		8.1
t_p		Maximum permitted normal operating conductor temperature	°C	2.5
U		Line voltage (voltage between lines)	V	2.2.1
U_0		Nominal line voltage to earth (for TN systems)	V	2.2.1
U_{oc}		Open circuit voltage at the distribution transformer	V	6.3.1
Z_a	*	Maximum adiabatic loop impedance	Ω	8.6
Z_{41}	*	The maximum value of earth fault loop impedance (Z_s) to provide disconnection within 0.4 s (Tables 41.2 and 41.3) or 5 s (Table 41.4 of BS 7671)	Ω	2.6
Z_x	*	Impedance of the transformer or supply	Ω	6.3.1
Z_D	*	Line impedance of the distribution cable	Ω	6.3.1
Z_{PEN}	*	Impedance of the PEN conductor	Ω	6.3.1
Z_1	*	Line impedance of the circuit conductor	Ω	6.3.1
Z_n	*	Neutral impedance of the circuit conductor	Ω	6.3.1
Z_2	*	Impedance of the circuit protective conductor	Ω	6.3.1
Z_e		That part of the earth fault loop impedance which is external to the installation	Ω	2.2.1
Z_s		Earth fault loop impedance	Ω	9.3
$\emptyset$		Phase angle	degrees	11

▼ **Table A.1**   Voltage factor c (the ratio between the equivalent voltage source and the nominal system voltage U_0 divided by $\sqrt{3}$)

Nominal voltage U_n	Voltage factor c for the calculation of:	
	maximum short-circuit currents c_{max} (Note 1)	minimum short-circuit currents c_{min}
Low voltage 100 V to 1 000 V (IEC 60038, Table I)	1.05 (Note 3) 1.10 (Note 4)	0.95
Medium voltage >1 kV to 35 kV (IEC 60038, Table III) **High voltage** (Note 2) >35 kV (IEC 60038, Table IV)	1.10	1.00

Notes:
1 $c_{max}U_n$ should not exceed the highest voltage U_m for equipment of power systems.
2 If no nominal voltage is defined $c_{max}U_n = U_m$ or $c_{min}U_n = 0.90 \times U_m$ should be applied.
3 For low voltage systems with a tolerance of +6%, for example systems renamed from 380 V to 400 V
4 For low voltage systems with a tolerance of +10%.

Note: The introduction of a voltage factor c is necessary for the following reasons:
▶ Voltage variation depending on time and place
▶ Changing of transformer taps
▶ Neglecting loads and capacitances by calculations
▶ The subtransient behaviour of generator and motors.

Standard final circuits

B

Tables B.1 to B.6 derive the standard circuits found in the IET *Electrician's Guide to the Building Regulations*. Users of the tables must confirm that the design assumptions are appropriate for the particular installation, particularly with respect to load and duration. Users must also check the calculations themselves. The approach taken is to calculate:

1   if the selected cable as installed can carry the load (equations 2.4.1 to 2.4.3)
2   the maximum length of cable that will meet the voltage drop limit (equations 2.5.1 to 2.5.3)
3   the maximum length of cable that will meet the fault protection requirement for shock protection (equations 2.6.1 and 2.6.2)
4   the maximum length of cable that will meet the adiabatic requirements (equations 2.8.1 and 2.8.2).

The maximum circuit length is then the smallest of 2) to 4) above.

The formulae used are given the equation references below (found in Chapter 2) in column 1 of the following tables:

$$(2.4.1) \quad I_t \geq \frac{I_n}{C_g\,C_a\,C_s\,C_d\,C_i\,C_f\,C_c}$$

$$(2.4.2) \quad I_t \geq \frac{20}{C_g\,C_a\,C_s\,C_d\,C_i\,C_f\,C_c}$$

$$(2.4.3) \quad I_t \geq \frac{I_b}{C_g\,C_a\,C_s\,C_d\,C_i\,C_f\,C_c}$$

$$(2.5.1) \quad L_{vd} = \frac{11.5 \times 1000}{I_b \times (mV/A/m) \times C_t} \quad \text{for radial circuits other than lighting}$$

$$(2.5.2) \quad L_{vd} = \frac{6.9 \times 1000}{(I_b/2) \times (mV/A/m) \times C_t} \quad \text{for lighting circuits with an evenly distributed load}$$

$$(2.5.3) \quad L_{vd} = \frac{4 \times 11.5 \times 1000}{I_b \times (mV/A/m) \times C_t} \quad \text{for ring circuits}$$

$$(2.5.4) \quad C_t = \frac{230 + t_p - \left(C_g^{\,2}\,C_a^{\,2}\,C_s^{\,2}\,C_d^{\,2} - \dfrac{I_b^{\,2}}{I_t^{\,2}}\right)(t_p - 30)}{230 + t_p}$$

Note: if t_p is 70 °C and $C_a = C_g = 1$, then

$$C_t = 1 - \frac{40}{300}\left(1 - \frac{I_b^2}{I_t^2}\right)$$

(2.6.1) $L_s = \dfrac{Z_{41} - Z_e}{(R''_1 + R''_2)\,C_r}$ for radial circuits

(2.6.2) $L_s = \dfrac{4(Z_{41} - Z_e)}{(R''_1 + R''_2)\,C_r}$ for ring circuits

(2.7.1) $L_{ss} = \dfrac{(Z_{41} - Z_e)}{2R_1\,C_r}$ for radial circuits

(2.7.2) $L_{ss} = \dfrac{4(Z_{41} - Z_e)}{2R_1\,C_r}$ for ring circuits

(2.8.1) $L_a = \dfrac{Z_a - Z_e}{(R''_1 + R''_2)\,C_r}$ for radial circuits

(2.8.2) $L_a = \dfrac{4(Z_a - Z_e)}{(R''_1 + R''_2)\,C_r}$ for ring circuits

where:

C_g is rating factor for grouping, see Table 4C1 of BS 7671 or F3 of the *On-Site Guide*

C_a is rating factor for ambient temperature, see Table 4B1 of BS 7671 or F1 of the *On-Site Guide*

C_s is rating factor for thermal resistivity of soil, see Table 4B3 of BS 7671

C_d is rating factor for depth of buried cable, see Table 4B4

C_i is rating factor for conductors surrounded by thermal insulation, see Regulation 523.9 of BS 7671 or Table F2 of the *On-Site Guide*

C_f is a rating factor applied when overload protection is being provided by an overcurrent device with a fusing factor greater than 1.45, e.g. $C_f = 0.725$ for semi-enclosed fuses to BS 3036.

C_c is a rating factor for buried circuits: 0.9 for cables buried in the ground requiring overload protection, otherwise is 1

C_r is rating factor for operating temperature (see Table F.3, Table I3 of the *On-Site Guide*)

C_t is a rating factor that can be applied if the load current is significantly less than I_z, the current-carrying capacity of the cable in the particular installation conditions. If C_t is taken as 1, any error will be on the safe side. This factor compensates for the temperature of the cable at the reduced current being less than the temperature at the maximum current information. See section 5.6

I_b is the design current

I_n is the rated (nominal) current or current setting of the protective device

I_t is the tabulated current-carrying capacity of a cable found in Appendix F of the *On-Site Guide* or Appendix 4 of BS 7671

I_z is the required current-carrying capacity of a cable for continuous service in its particular installed condition

L   is the length of cable in the circuit

L_a is the maximum cable length if the adiabatic equation of Regulation 543.1.3 is to be met

L_s is the maximum length of cable in the circuit if the disconnection requirements in the event of a fault are to be met

L_{ss} is the maximum cable length for short-circuit protection to be ensured

L_{vd} is cable length if the voltage drop limit is not to be exceeded

(mV/A/m)   is the voltage drop per ampere per metre from Appendix 4

R''_1 is resistance of the line conductor per metre (see Table F.1, Table I1 of the *On-Site Guide*)

R''_2 is resistance of the protective conductor per metre (see Table F.1, Table I1 of the *On-Site Guide*)

Z_{41} is maximum earth fault loop impedance given by the appropriate Table 41.2, 41.3 or 41.4

Z_a is maximum adiabatic loop impedance, see Chapter 8 for explanation and tables of values

Z_e is earth fault loop impedance external to the circuit – in this section it is assumed to be that of the supply, i.e. 0.8 or 0.35 Ω.

References used in the following tables:

F.1   is to Table F.1 in Appendix F of this Guide (*Electrical Installation Design Guide*)

8.6   is to Table 8.6 of this Guide

4D5 (R) etc. is to Table 4D5 of the *IET Wiring Regulations*

NP   is Not Permitted

41.2 (R)   is to Table 41.2 of the *IET Wiring Regulations*.

▼ **Table B.1**   Radial final socket-outlet circuit (Note 7)

Equation or table	Element calculated in table	Device type								
		Fuse BS				c.b. type				
		88-3	3036	3036	88-2	1	2	B	3 or C	D
	I_n (A)	20	20	20	20	20	20	20	20	20
	I_b (A) average [9]	16.5	16.5	16.5	16.5	16.5	16.5	16.5	16.5	16.5
2.4.1	Required I_t (A) [1] ≥	20	27.60	27.60	20	20	20	20	20	20
	Cable L/PE [2]	2.5/1.5	2.5/1.5	4.0/1.5 [8]	2.5/1.5	2.5/1.5	2.5/1.5	2.5/1.5	2.5/1.5	2.5/1.5
4D5 (R)	Installation method [3]	A	C	100	A	A	A	A	A	A
	Cable I_t (A)	20	27	27	20	20	20	20	20	20
2.5.4	C_t	0.96	0.92	0.92	0.96	0.96	0.96	0.96	0.96	0.96
4D5 (R)	(mV/A/m)	18	18	11	18	18	18	18	18	18
2.5.1	L_{VD} (m)	40.1	42.3	69.1	40.1	40.1	40.1	40.1	40.1	40.1
	Shock [4]									
41.2, 41.3 (R)	Z_{41} (Ω)	2.04	1.77	1.77	1.77	2.88	1.64	2.30	1.15	0.57
F.1	$R''_1 + R''_2$ (mΩ/m)	19.51	19.51	16.71	19.51	19.51	19.51	19.51	19.51	19.51
F.3	C_r	1.2	1.2	1.2	1.2	1.2	1.2	1.2	1.2	1.2
	Z_e, 0.8 (Ω)									
2.6.1	L_s, 0.4 s (m)	52.9	41.4	48.4	41.4	88.6	36.0	64.1	14.9	NP
	Z_e, 0.35 (Ω)									
	L_s, 0.4 s (m)	72.2	60.4	70.8	60.7	107.9	55.2	83.2	34.2	10.7
	Adiabatic									
8.6 to 8.8	Z_a (Ω)	3.70	4.3	4.3	3.1					
2.8.1	L_a, 0.35 (m)	123	165	340	116	Note 5	Note 5	Note 5	Note 5	Note 5
	L_a, 0.8 (m)	143	145	270	96					
	Short-circuit									
F.1	$2 R''_1$	14.82	14.82	9.22	14.82	14.82	14.82	14.82	14.82	14.82
	Z_e, 0.8 (Ω)									
2.7.1	L_{ss}, 5 s (m)	145	170	273	121	116	47	84	19	NP
	Z_e, 0.35 (Ω)									
	L_{ss}, 5 s (m)	180	195	314	145	142	72	109	40	12
Maximum lengths with no RCD [6]	L, 0.8 (m)	40	41	48	40	40	36	40	15	NP
	L, 0.35 (m)	40	42	69	40	40	40	40	34	9.6
Maximum lengths with RCD [6]	L, 0.8 (m)	40	41	69	40	40	40	40	19	NP
	L, 0.35 (m)	40	42	69	40	40	40	40	40	12

Notes to Table B.1:

1 Overcurrent protection is required.

2 70 °C thermoplastic flat cable with protective conductor per Table 4D5 of BS 7671.

3 Circuits have been designed for the installation reference method listed (generally A). The circuit may be used in less onerous conditions. For example, a circuit designed for installation reference method A may also be used for reference methods 100, 102, B and C.

4 Disconnection required in 0.4 second for TN installations.

5 See Tables 8.4 and 8.5 for minimum protective conductor sizes.

6 L 0.8 signifies the maximum length for installation method A and external supply impedance $Z_e = 0.8\ \Omega$.

7 Socket circuits will in most circumstances be required to be additionally protected by a 30 mA RCD (Regulation 411.3.3).

8 A rule-based selection may require a 4 mm² cable, however a designer may exercise judgement and select a 2.5 mm² cable.

9 Design load current based on 13 A at the extremity and the balance to the device rating (20 A) evenly distributed $I_t = (13 + 20)/2 = 16.5$.

▼ **Table B.2**   Ring final circuit supplying socket-outlets (Note 7)

Equation or table	Element calculated in table	Device type							
		Fuse BS			c.b. type				
		88-3	3036	88-2	1	2	B	3 or C	D
	I_n (A)	32	30	32	30	30	32	32	32
	I_b (A) [1, 8]	26	25	26	25	25	26	26	26
433.1.103 (R)	Required I_t (A) [1] ≥	20	20	20	20	20	20	20	20
4D5 (R)	Cable L/PE [2]	2.5/1.5	2.5/1.5	2.5/1.5	2.5/1.5	2.5/1.5	2.5/1.5	2.5/1.5	2.5/1.5
	Installation method [3]	A	A	A	A	A	A	A	A
	Cable I_t (A)	20	20	20	20	20	20	20	20
2.5.4	C_t	0.92	0.92	0.92	0.92	0.92	0.92	0.92	0.92
4D5 (R)	(mV/A/m)	18	18	18	18	18	18	18	18
2.5.3	L_{VD} (m)	111	111	111	111	111	106	106	106
	Shock [4]								
41.2, 41.3 (R)	Z_{41} (Ω)	0.96	1.09	1.04	1.92	1.10	1.44	0.72	0.36
F.1	$R''_1 + R''_2$ (mΩ/m)	19.50	19.50	19.50	19.50	19.50	19.50	19.50	19.50
F.3	C_r	1.2	1.2	1.2	1.2	1.2	1.2	1.2	1.2
	Z_{e}, 0.8 (Ω)								
2.6.2	L_S (m)	27	49	41	190	50	108	NP	NP
	Z_{e}, 0.35 (Ω)								
	L_S (m)	103	126	117	267	127	185	63	1.6
	Adiabatic								
8.6 to 8.8	Z_a (Ω)	1.1	2.49	1.34					
2.8.2	L_a, 0.35 (m)	128	365	169	Note 5	Note 5	Note 5	Note 5	Note 5
	L_a, 0.8 (m)	51	289	92					
	Short-circuit								
	Z_{e}, 0.8 (Ω)								
2.7.2	L_{ss}, 5 s (m)	188	413	233	255	66	143	NP	NP
	Z_{e}, 0.35 (Ω)								
	L_{ss}, 5 s (m)	290	515	335	352	167	244	83	2
Maximum lengths with no RCD [6]	L, 0.8 (m)	27	49	41	111	50	106	NP	NP
	L, 0.35 (m)	103	111	111	111	111	106	63	1.6
Maximum lengths with RCD [6]	L, 0.8 (m)	111	111	111	111	66	106	NP	NP
	L, 0.35 (m)	111	111	111	111	111	106	83	2

Notes to Table B.2:

1 Regulation 433.1.103 requires $I_Z \geq 20$ A. Overcurrent protection is not required; this is significant for rewirable fuses.

2 70 °C thermoplastic flat cable with protective conductor per Table 4D5 of BS 7671.

3 Circuits have been designed for the installation reference method listed (generally A). The circuit may be used in less onerous conditions. For example, a circuit designed for installation reference method A may also be used for reference methods 100, 102, B and C.

4 Disconnection required in 0.4 second for TN installations.

5 See Tables 8.4 and 8.5 for minimum protective conductor sizes.

6 L 0.8 signifies the maximum length for installation method A and external supply impedance $Z_e = 0.8\ \Omega$.

7 Socket circuits will in most circumstances be required to be additionally protected by a 30 mA RCD (Regulation 411.3.3).

8 Design load current based on 20 A load at the extremity and the balance to device rating (10 or 12 A) evenly balanced. This equates to 25 A for a 30 A device. 20 A is the maximum current socket-outlets to BS 1363 are required to handle.

▼ **Table B.3** Cooker circuit protected by a 30/32 A overcurrent device (Note 7)

Equation or table	Element calculated in table	Device type								
		Fuse BS				c.b. type				
		88-3	3036	3036	88-2	1	2	B	3 or C	D
	I_n (A) [1]	32	30	30	32	30	30	32	30	32
	I_b (A)	30	30	30	30	30	30	30	30	30
2.4.1	Required I_t (A) ≥	30	41.4	41.4	30	30	30	30	30	30
	Cable L/PE [2]	6/2.5	6/2.5	10/4	6/2.5	6/2.5	6/2.5	6/2.5	6/2.5	6/2.5
	Installation method [3]	A	C	A	A	A	A	A	A	A
	Cable I_t (A)	32	46	44	32	32	32	32	32	32
2.5.4	C_t	0.98	0.92	0.93	0.98	0.98	0.98	0.98	0.98	0.98
4D5 (R)	(mV/A/m)	7.30	7.30	6.44	7.30	7.30	7.30	7.30	7.30	7.30
2.5.1	L_{VD} (m)	53.4	57	93	53.4	53.4	53.4	53.4	53.4	53.4
	Shock [4]									
41.2, 41.3 (R)	Z_{41} (Ω)	0.96	1.09	1.09	1.04	1.92	1.10	1.44	0.72	0.36
F.1	$R''_1 + R''_2$ (mΩ/m)	10.49	10.49	6.44	10.49	10.49	10.49	10.49	10.49	10.49
F.3	C_r	1.2	1.2	1.2	1.2	1.2	1.2	1.2	1.2	1.2
	Z_e, 0.8 (Ω)									
2.6.1	L_S (m)	12.6	23	37.5	19.1	88	23	30	NP	NP
	Z_e, 0.35 (Ω)									
	L_S (m)	48.3	56.8	95.8	54.8	124	59	86	29	1
	Adiabatic									
8.6 to 8.8	Z_a (Ω)	1.68	3.16	3.55	1.9					
2.8.1	L_a, 0.35 (m)	105	223	413	123	Note 5	Note 5	Note 5	Note 5	Note 5
	L_a, 0.8 (m)	69	187	355	87					
	Short-circuit									
	Z_e, 0.8 (Ω)									
2.7.1	L_{ss}, 5 s (m)	113	248	418	140	151	39	86	NP	NP
	Z_e, 0.35 (Ω)									
	L_{ss}, 5 s (m)	174	309	521	201	211	100	147	49	1
Maximum lengths with no RCD [6]	L, 0.8 (m)	12	23	37	19	53	23	53	NP	NP
	L, 0.35 (m)	48	57	93	53	53	53	53	29	1
Maximum lengths with RCD [6]	L, 0.8 (m)	53	57	93	53	53	39	53	53 (NP)	53 (NP)
	L, 0.35 (m)	53	57	93	53	53	53	53	53 (49)	53 (1)

Notes to Table B.3:

1. Using Table 3.1, assume a cooker rated at up to 14.4 kW at 240 V and assuming a socket is incorporated in the cooker control unit, then $I_b = 10 + 0.3 (60 - 10) + 5 = 30$ A.

2. 70 °C thermoplastic flat cable with protective conductor per Table 4D5 of BS 7671.

3. Circuits have been designed for the installation reference method listed (generally A). The circuit may be used in less onerous conditions. For example, a circuit designed for installation reference method A may also be used for reference methods 100, 102, B and C.

4. Disconnection required in 0.4 second for TN installations for 30 and 32 A circuits.

5. See Tables 8.4 and 8.5 for minimum protective conductor sizes.

6. L 0.8 signifies the maximum length for installation method A and external supply impedance $Z_e = 0.8$ Ω.

7. Cooker circuits with a cooker control unit incorporating a socket-outlet will be required to be additionally protected by a 30 mA RCD (Regulation 411.3.3).

B

▼ **Table B.4** Shower circuits for up to 7.21 kW and 9.6 kW ratings at 240 V (Note 7)

Equation or table	Element calculated in table	Device type — Fuse BS: 88-3	88-3	3036	3036	88-2	88-2	c.b. type: 1	1	2	2	B	B	3 or C	3 or C	D	D
	I_n (A) [1]	32	40	45	45	30	40	32	40	32	40	32	40	32	40	32	40
	I_b (A)	30	40	30	40	30	40	30	40	30	40	30	40	30	40	30	40
2.4.1	Rating kW	7.20	9.60	7.20	9.60	7.20	9.60	7.20	9.60	7.20	9.60	7.20	9.60	7.20	9.60	7.20	9.60
	Required I_t (A) ≥	30[1]	40[1]	40[1]	40[1]	30[1]	40	30	40	30	40	30	40	30	40	30	40
4D5 (R)	Cable L/PE [2]	6/2.5	10/4	10/4.	10/4	6/2.5	10/4	6/2.5	10/4	6/2.5	10/4	6/2.5	10/4	6/2.5	10/4	6/2.5	10/4
	Installation method [3]	A	A	A	A	A	A	A	A	A	A	A	A	A	A	A	A
	Cable I_t (A)	32	44	44	44	32	44	32	44	32	44	32	44	32	44	32	44
2.5.4	C_t	0.98	0.98	0.98	0.98	0.98	0.98	0.98	0.98	0.98	0.98	0.98	0.98	0.98	0.98	0.98	0.98
4D5 (R)	(mV/A/m)	7.30	4.40	4.40	4.40	7.30	4.40	7.30	4.40	7.30	4.40	7.30	4.40	7.30	4.40	7.30	4.40
2.5.1	L_{VD} (m)	53.4	66.9	66.9	66.9	53.4	66.9	53.4	66.9	53.4	66.9	53.4	66.9	53.4	66.9	53.4	66.9
	Shock [4]																
41.2, 41.3 (R)	Z_{41} (Ω)	0.96	1.04	1.04	1.09	1.59	1.04	1.92	1.35	1.44	1.10	0.82	1.15	0.72	0.58	0.36	0.29
F.1	$R_1' + R_2'$ (mΩ/m)	10.49	6.44	6.44	6.44	10.49	6.44	10.49	6.44	10.49	6.44	10.49	6.44	10.49	6.44	10.49	6.44
F.3	C_r	1.2	1.2	1.2	1.2	1.2	1.2	1.2	1.2	1.2	1.2	1.2	1.2	1.2	1.2	1.2	1.2
2.6.1	Z_e, 0.8 (Ω) L_s (m)	12.5	31	23	102	88	19	82	23	50	45	3	NP	NP	NP	NP	NP
	Z_e, 0.35 (Ω) L_s (m)	48	89	58	160	124	54	140	59	86	103	61	45	29	29	1	NP

continues

Calculations for Electricians and Designers
© The Institution of Engineering and Technology

▼ **Table B.4** *continued*

Equation or table	Element calculated in table	Device type								
		Fuse BS				**c.b. type**				
		88-3	3036	88-2	1	1	2	B	3 or C	D
Note 3	**Adiabatic**									
8.6 to 8.8	Z_a (Ω)	1.64	1.1	3.16	1.92	1.92	1.44	Note 5	Note 5	Note 5
	L_a, 0.35 (m)	132	97	223	202	124	140	Note 5	Note 5	Note 5
2.8.1	L_a, 0.8 (m)	66	38	187	144	88	82	Note 5	Note 5	Note 5
	Short-circuit									
2.7.1	Z_e, 0.8 (Ω)									
	L_{ss}, 5 s (m)	113	54	248	179	39	5	79	NP	NP
	Z_e, 0.35 (Ω)									
	L_{ss}, 5 s (m)	174	170	309	282	59	107	182	51	1
Maximum lengths with no RCD [6]	L, 0.8 (m)	53	23	66	19	23	3	50	NP	NP
	L, 0.35 (m)	67 (54)	53	66	53	53	61	45	49	1
Maximum lengths with RCD [6]	L, 0.8 (m)	53	53	67	53	39	5	53	NP	NP
	L, 0.35 (m)	67	53	67	53	53	66	66	51	1

Notes:

1 Overcurrent protection is not required as shower loads are fixed; this is significant for rewirable fuses.

2 70 °C thermoplastic flat cable with protective conductor per Table 4D5 of BS 7671.

3 Circuits have been designed for the installation reference method listed (generally A). The circuit may be used in less onerous conditions. For example, a circuit designed for installation reference method A may also be used for reference methods 100, 102, B and C.

4 Disconnection required in 0.4 second for TN installations for 30 and 32 A circuits, 5 seconds for 40 and 45 A.

5 See Tables 8.4 and 8.5 for minimum protective conductor sizes.

6 L 0.8 signifies the maximum length for installation method A and external supply impedance Z_e = 0.8 Ω.

7 Shower circuits are required to be additionally protected by a 30 mA RCD (Regulation 701.411.3.3).

▼ Table B.5 Lighting circuits (take note of device ratings I_n and Note 1)

Equation or table	Element calculated in table	Device type							
		Fuse BS			c.b. type				
		88-3	3036	88-2	1	2	B	3 or C	D
	I_n (A) [1]	5	5	10	10	10	10	10	6
	I_b (A) [1]	5	5	5	5	5	5	6	5
2.4.1	Required I_t (A) ≥	5	6.9	10	10	10	10	6	6
	Cable L/PE [2]	1.5/1	1.5/1	1.5/1	1.5/1	1.5/1	1.5/1	1.5/1	1.5/1
	Installation method [3]	103	103	103	103	103	103	103	103
4D5	Cable I_t (A)	10	10	10	10	10	10	10	10
2.5.4	C_t	0.88	0.88	0.88	0.88	0.88	0.88	0.88	0.88
4D5	(mV/A/m)	29	29	29	29	29	29	29	29
2.5.2	L_{VD} M6 (m) [5]	108	108	108	108	108	108	108	108
	Shock [4]								
41.2, 41.3 (R)	Z_s, 5 s (Ω)	10.45	9.58	4.89	5.75	3.29	4.60	3.83	1.92
F.1	$R''_1 + R''_2$ (mΩ/m)	30.20	30.20	30.20	30.20	30.20	30.20	30.20	30.20
F.3	C_r	1.2	1.2	1.2	1.2	1.2	1.2	1.2	1.2
	Z_e, 0.8 (Ω)								
2.6.1	L_s (m)	266	242	112	136	68	104	87	30
	Z_e, 0.35 (Ω)								
	L_s (m)	278	254	125	149	81	117	100	43
	Adiabatic								
8.6 to 8.8	Z_a (Ω)	21	23	8.5					
2.8.1	L_a, 0.35 (m)	570	625	224	Note 5	Note 5	Note 5	Note 5	Note 5
	L_a, 0.8 (m)	557	612	212					
	Short-circuit								
F.1	$2R_1$	24.2	24.2	24.2					
	Z_e, 0.8 (Ω)								
2.7.1	L_{ss} (m)	711	764	265	170	85	130	104	38
	Z_e, 0.35 (Ω)								
	L_{ss}, (m)	795	780	280	186	101	144	119	53
Maximum lengths with no RCD [6]	L, 0.8 (m)	108	108	108	108	68	104	83	30
	L, 0.35 (m)	108	108	108	108	81	108	96	43
Maximum lengths with RCD [6]	L, 0.8 (m)	108	108	108	108	108	108	104	38
	L, 0.35 (m)	108	108	108	108	108	108	108	53

Notes to Table B.5:

1 Load has been presumed to be 5 A evenly distributed even when 10 A device is installed (10 A selected for type B c.bs to avoid unwanted tripping).

2 70 °C thermoplastic flat cable with protective conductor per Table 4D5 of BS 7671.

3 Circuits have been designed for the installation method listed. The circuit may be used in less onerous conditions. For example, a circuit designed for installation method 103 may also be used for reference methods 101, A, 100, 102, B and C.

4 Disconnection required in 0.4 second for TN installations.

5 See Tables 8.4 and 8.5 for minimum protective conductor sizes.

6 L 0.8 signifies the maximum length for installation method A and external supply impedance $Z_e = 0.8\ \Omega$.

▼ **Table B.6** Immersion and storage heater circuit

Equation or table	Element calculated in table	Device type							
		Fuse BS			c.b. type				
		88-3	3036	88-2	1	2	B	3 or C	D
	I_n (A)	16	15	16	16	16	16	16	16
	I_b (A)[1]	12.5	12.5	12.5	12.5	12.5	12.5	12.5	12.5
2.4.1	Required I_t (A) ≥	16	17.24	16	16	16	16	16	16
	Cable L/PE[2]	2.5/1.5	2.5/1.5	2.5/1.5	2.5/1.5	2.5/1.5	2.5/1.5	2.5/1.5	2.5/1.5
	Installation method[3]	101	101	101	101	101	101	101	101
4D5	Cable I_t (A)	13.5	13.5	13.5	13.5	13.5	13.5	13.5	13.5
2.5.4	C_t	0.94	0.94	0.94	0.94	0.94	0.94	0.94	0.94
4D5	(mV/A/m)	18	18	18	18	18	18	18	18
2.5.1	L_{VD} (m)	54	54	54	54	54	54	54	54
	Shock[4]								
41.2, 41.3 (R)	Z_{41} (Ω)	2.42	2.55	2.56	3.59	2.05	2.87	1.44	0.72
F.1	$R''_1 + R''_2$ (mΩ/m)	19.51	19.51	19.50	19.50	19.50	19.50	19.50	19.50
F.3	C_r	1.2	1.2	1.2	1.2	1.2	1.2	1.2	1.2
	Z_e, 0.8 (Ω)								
2.6.1	L_s (m)	69	75	75	119	53	88	27	NP
	Z_e, 0.35 (Ω)								
	L_s (m)	88	94	94	138	72	107	46	16
	Adiabatic								
8.6 to 8.8	Z_a (Ω)	4.6	6.38	4.79					
	C_r	1.2	1.2	1.2	Note 5	Note 5	Note 5	Note 5	Note 5
2.8.1	L_a, 0.35 (m)	181	257	191					
	L_a, 0.8 (m)	162	238	170					
	Short-circuit								
	Z_e, 0.8 (Ω)								
2.7.1	L_{ss} (m)	186	339	190	157	70	116	35	NP
	Z_e, 0.35 (Ω)								
	L_{ss}, (m)	239	255	150	182	95	142	61	20
Maximum lengths with no RCD[6]	L, 0.8 (m)	54	54	54	54	53	54	27	NP
	L, 0.35 (m)	54	54	54	54	54	54	46	16
Maximum lengths with RCD[6]	L, 0.8 (m)	54	54	54	54	54	54	35	NP
	L, 0.35 (m)	54	54	54	54	54	54	54	20

Notes to Table B6:

1. Load has been presumed to be 12.5 A (3 kW at 240 V).
2. 70 °C thermoplastic flat cable with protective conductor per Table 4D5 of BS 7671.
3. Circuits have been designed for the installation method listed. The circuit may be used in less onerous conditions. For example, a circuit designed for installation method 101 may also be used for reference methods A, 100, 102, B and C.
4. Disconnection required in 0.4 second for TN installations.
5. See Tables 8.4 and 8.5 for minimum protective conductor sizes.
6. L 0.8 signifies the maximum length for installation method A and external supply impedance $Z_e = 0.8\ \Omega$.

B

Avoidance of unintentional operation of circuit-breakers

C

(The following typical manufacturer's guidance is reproduced with the kind permission of Hager.)

The unintentional operation of circuit-breakers is most commonly known as nuisance tripping, and care must be taken in the selection of circuit-breakers to prevent their unintentional operation.

Regulation 533.2.1 states that:

> The rated current (or current setting) of the protective device shall be chosen in accordance with Regulation 433.1. In certain cases, to avoid unintentional operation, the peak current values of the loads may have to be taken into consideration.

The peak current values of a load are the starting characteristics, which may include inrush current. The load characteristics should be compared with the minimum tripping current of the circuit-breaker, see Figure C.1.

▼ **Figure C.1**
Circuit-breaker characteristic

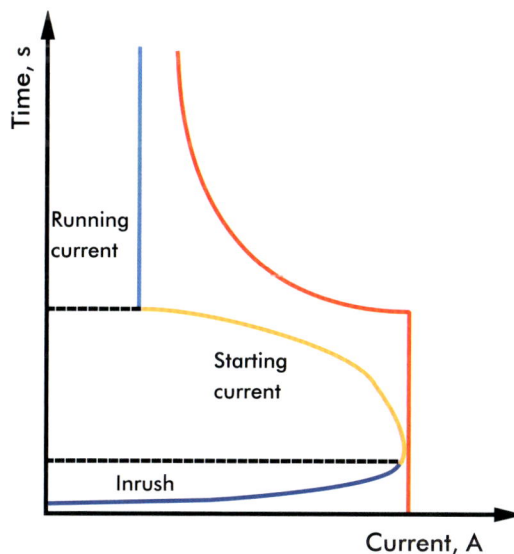

Inrush peak load current should be compared with the minimum peak tripping current of the circuit-breaker to ensure that unwanted tripping is avoided. The inrush peak current value is obtained from the manufacturer of the load equipment, and the minimum peak tripping current calculated by multiplying I_n by the lower multiple of the instantaneous trip and then by 1.414 (crest factor).

▼ **Table C.1**   Circuit-breaker minimum peak tripping current (A)

Circuit-breaker type	Circuit-breaker rated current								
	6 A	10 A	16 A	20 A	25 A	32 A	40 A	50 A	63 A
B	26	43	68	85	106	136	170	212	268
C	43	71	113	142	177	223	283	354	446
D	85	142	226	283	354	453	566	707	891

Circuit-breaker frame	Minimum peak tripping current		
	I_n	min	max
H125D	–	–	–
H125H	16–125	–	1131
H250N	160	905	1810
	200	1131	2262
	250	1414	2828
H400N	320	1810	3620
	400	2262	4525
H630N	500	2828	5656
	630	3563	7126
H800N	800	4525	9050

For other settings use: $(I_n \times \text{magnetic setting} - 20\%) \times 1.414$

C.1   Lighting circuit applications

For the protection of lighting circuits the designer must select the circuit-breaker with the lowest instantaneous trip current compatible with the inrush currents likely to develop in the circuit. High-frequency (HF) ballasts are often singled out for their high inrush currents but they do not differ widely from conventional 50 Hz versions. The highest value is reached when the ballast is switched on at the moment the mains sine wave passes through zero. The HF system is a 'rapid start' arrangement, in which all lamps start at the same time. Therefore the total inrush current of a lighting circuit incorporating HF ballasts exceeds the usual values of a conventional 50 Hz system. Where multiple ballasts are used in lighting schemes, the peak current increases proportionally. Mains circuit impedance will reduce the peak current but will not affect the pulse time. The problem facing the installation designer in selecting the correct circuit-breaker is that the surge characteristics of high-frequency ballasts vary from manufacturer to manufacturer. Some may be as low as 12 A with a pulse time of 3 milliseconds and some as high as 35 A with a pulse time of 1 millisecond. Therefore it is important to obtain the expected inrush current of the equipment from the manufacturer in order to find out how many HF ballasts can safely be supplied from one circuit-breaker without the risk of nuisance tripping. This information can then be divided into the minimum peak tripping current of the circuit-breaker.

Example

How many HF ballasts, each having an expected inrush of 20 A, can be supplied by a 16 A type C circuit-breaker?

From Table C.1, 16 A type C, we have a minimum peak tripping current of 113 A.

Therefore 113/20 = 5.65

Hence, a maximum of 5 ballasts can be supplied by a 16 A type C circuit-breaker.

c

Further cable calculations

D

433 ## D.1 Cable life

British Cables Association

The advice of the British Cables Association is: 'Estimating the life of a cable can only be approximate.'

There is no definitive or simple calculation method that can be used to determine the life expectancy of a fixed wiring cable.

Many factors determine the life of a cable, including for example:

▶ mechanical damage
▶ presence of water
▶ chemical contamination
▶ solar or infrared radiation
▶ number of overloads
▶ number of short-circuits
▶ effects of harmonics
▶ temperature at terminations
▶ temperature of the cable.

Provided an appropriate good quality cable has been selected taking into account operating temperature, installation conditions, equipment it is being attached to, and in accordance with appropriate regulations, it should meet or exceed its design life.

The design life of good quality fixed wiring cables is in excess of 20 years, when appropriately selected and installed. This design life has been assessed on a maximum loading – that is, the cable running at its maximum conductor operating temperature for 24 hours a day and 365 days a year.

If an installation is not fully loaded all the time, the expected life of the cable would be greater than the design life of the cable.

There are many instances of good quality fixed wiring cables operating for in excess of 40 years; this is mainly due to the cables not being abused and being lightly or periodically loaded.

D.2 Temperatures (core and sheath) (Effect of load current on conductor operating temperature)

The conductor operating temperature at other than the full load current can be determined from the equation:

$$\frac{I_b}{I_t} = \sqrt{\frac{t_b - t_a}{t_p - t_o}} \qquad \text{hence } t_b = \frac{I_b^2}{I_t^2}(t_p - t_o) + t_a$$

where:

I_b = load current resulting in a conductor temperature t_b at an ambient t_a

I_t = tabulated current rating in Appendix 4 of BS 7671, resulting in a conductor temperature t_p at ambient t_o.

This relationship assumes that temperature rise of a cable is proportional to the square of the current.

523.7, Appx 10

D.3 Inductance of cables in parallel

Consider 3 x 300 mm² aluminium conductor cables, non-armoured, installed as per Reference Method F, with the spacing being considered as either one cable diameter or two cable diameters. BS 7671 Table 4J1B column 9 gives the impedance of the cables at 90 °C in mV/A/m (or mΩ/m) as:

r	x	$z = \sqrt{r^2 + x^2}$
0.22	0.3	0.37

Figure D.1 gives that, if the actual spacing is two cable diameters, the additional reactance is 0.1 mΩ/m. The impedance of the cable now becomes:

r	x	$z = \sqrt{r^2 + x^2}$
0.22	0.4	0.46

▼ **Figure D.1**
Additional reactance for non-armoured single-core cables with wider spacing than one cable diameter

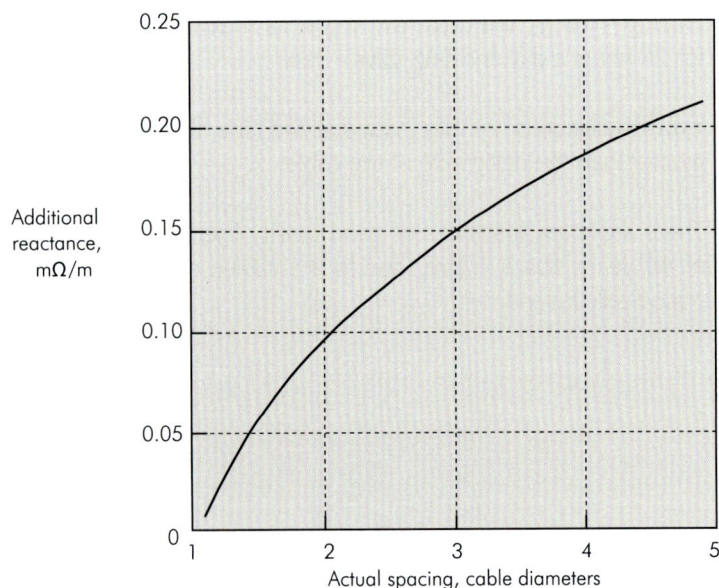

The voltage drop tables make reference to units of (mV/A/m) for r, x and z.

Note that cables with uneven spacing will have uneven balance of load between them. There are certain cable arrangements that provide reasonable current sharing (examples being given in Figure 9.8 of the *Commentary*).

Other arrangements may provide acceptable current sharing, but consideration will have to be given to the different reactances, and the values must be calculated.

523.100 D.4 Calculation of sheath voltages

▼ **Figure D.2**
Voltages on single-core sheathed cables

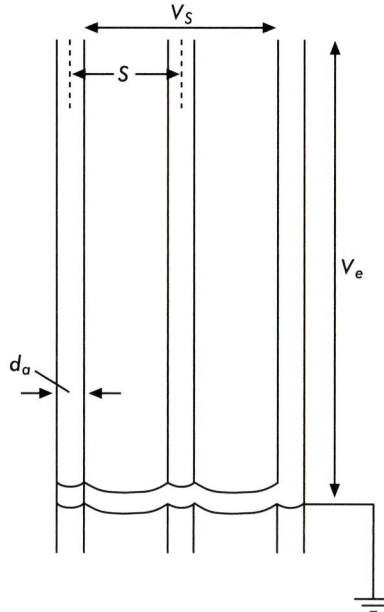

Sheath voltages in single point bonded systems operating at 50 Hz may be calculated using the following equations:

Single-phase	$V_s = 2XI$	V/km
	$V_e = XI$	V/km
Three-phase, trefoil	$V_s = \sqrt{3}XI$	V/km
	$V_e = XI$	V/km
Three-phase, flat	$V_s = \sqrt{3}I\,(X + X_m)$	V/km
	$V_e = I\sqrt{(X^2 + XX_m + X_m^2)}$	V/km

where:

X_m = 0.0433 Ω/km
X = 0.0628 $\log_e(2\,S/d_a)$ Ω/km
S = spacing between cable centres in mm
d_a = mean sheath or armour diameter in mm
V_s = voltage between sheaths, between outers for three flat arrangements
V_e = voltage sheath to earth, outer to earth for three flat arrangements
I = line current.

▼ **Table D.1** Sheath to sheath voltage for cables to BS 5467 (mV/A/m). Single point bonding of sheath or armour; voltages between sheaths or armouring of outer conductors.

Nominal size (mm²)	Single-phase			Three-phase			
	Touching (mV)	Spaced 2 x D_e (mV)	Spaced 5 x D_e (mV)	Trefoil (mV)	Flat touching (mV)	Spaced 2 x D_e (mV)	Spaced 5 x D_e (mV)
50.0	0.119	0.206	0.322	0.103	0.179	0.254	0.354
70.0	0.117	0.204	0.320	0.101	0.177	0.253	0.353
95.0	0.115	0.203	0.318	0.100	0.175	0.251	0.351
120.0	0.113	0.200	0.316	0.098	0.173	0.249	0.349
150.0	0.113	0.200	0.316	0.098	0.173	0.249	0.349
185.0	0.111	0.199	0.314	0.096	0.172	0.248	0.348
240.0	0.109	0.196	0.312	0.094	0.170	0.246	0.346
300.0	0.108	0.195	0.311	0.094	0.169	0.245	0.345
400.0	0.108	0.195	0.310	0.093	0.169	0.244	0.344
500.0	0.106	0.194	0.309	0.092	0.168	0.243	0.343
630.0	0.105	0.192	0.308	0.091	0.166	0.242	0.342
800.0	0.105	0.192	0.308	0.091	0.167	0.242	0.342
1000.0	0.104	0.191	0.307	0.090	0.166	0.241	0.341

Notes:

1 Voltages given for three-phase flat touching and spaced cables refer to the outer cables.
2 D_e is the overall cable diameter.

▼ **Table D.2** Sheath or armour voltage to earth for cables to BS 5467, BS 6346 and BS 6480 (mV/A/m). Single point bonding of sheath or armour; voltages between sheaths or armouring of outer conductors.

Nominal size (mm²)	Single-phase			Three-phase			
	Touching (mV)	Spaced 2 x D_e (mV)	Spaced 5 x D_e (mV)	Trefoil (mV)	Flat touching (mV)	Spaced 2 x D_e (mV)	Spaced 5 x D_e (mV)
50.0	0.060	0.103	0.161	0.060	0.090	0.131	0.187
70.0	0.059	0.102	0.160	0.059	0.089	0.130	0.186
95.0	0.058	0.101	0.159	0.058	0.088	0.129	0.185
120.0	0.056	0.100	0.158	0.056	0.087	0.128	0.184
150.0	0.056	0.100	0.158	0.056	0.087	0.128	0.183
185.0	0.056	0.099	0.157	0.056	0.086	0.127	0.183
240.0	0.055	0.098	0.156	0.055	0.085	0.126	0.182
300.0	0.054	0.098	0.155	0.054	0.085	0.125	0.181
400.0	0.054	0.097	0.155	0.054	0.085	0.125	0.181
500.0	0.053	0.097	0.155	0.053	0.084	0.125	0.180
630.0	0.053	0.096	0.154	0.053	0.083	0.124	0.180
800.0	0.053	0.096	0.154	0.053	0.083	0.124	0.180
1000.0	0.052	0.096	0.153	0.052	0.083	0.123	0.179

Notes:
1 Voltages given for three-phase flat touching and spaced cables refer to the outer cables.
2 D_e is the overall cable diameter.

Figure D.2 shows sheath to sheath and sheath to earth voltages.

Tables D.1 and D.2 tabulate voltages calculated using the formulae above for cables to BS 5467, BS 6346 and BS 6480.

Example

Consider 2 x 630 mm² aluminium conductor cables per phase supplying the load from a 1 MVA transformer. The full-load line current per cable is of the order of 700 A.

Assuming a two diameter spacing, the sheath or armour to earth voltage given by Table D.2 is 0.124 x 700/1000 V/m = 0.087 V/m.

Cable length of the order of 250 m would be necessary to generate 25 V to earth under full-load conditions. Such transformer cables are rarely longer than 15 m.

Symbols from CENELEC report R064-003

E

I_b — Design current of the circuit being considered (A) (IEV 826-05-04)

I_z — Current-carrying capacity of a conductor (A) (IEV 826-05-05)

I_f — Fault current (A)

I_{nc} — Rated current of busbar trunking system, in an ambient temperature of 30 °C (A)

(I^2t) — Thermal stress capacity of line, neutral or PE (PEN) conductor given in general for one second (A^2.s) (IEV 447-07-17 and EN 60439-2, 4.3)

L_1 — Route length (m)

L_2 — Length of busbar trunking system (m)

R_N — Resistance of the neutral conductor upstream of the circuit being considered

R_{PE} — Resistance of the protective conductor from the main equipotential bonding to the origin of the circuit being considered

R_{PEN} — Resistance of the PEN conductor from the main equipotential bonding to the origin of the circuit being considered

R_Q — Resistance upstream of the source

R_T — Resistance of the source

$R_{b1\,ph}$ — Mean ohmic resistance of busbar trunking system (BTS) per metre, per phase, at rated current I_{nc} at the steady-state operating temperature

R_{b0} — Resistive term of mean line-line, line-neutral or line-PE (-PEN) BTS loop impedance, at 20 °C

R_{b1} — Resistive term of mean line-line, line-neutral or line-PC (-PEN) BTS loop impedance at rated current I_{nc}, at the steady-state operating temperature

R_{b2} — Resistive term of mean line-line, line-neutral or line-PE (-PEN) BTS loop impedance at the mean temperature between the operating temperature at rated current I_{nc}, and the maximum temperature under short-circuit conditions

R_U — Resistance of the line conductor upstream of the circuit being considered

S — Cross-sectional area of conductor

S_N — Cross-sectional area of neutral conductor

S_{PE} — Cross-sectional area of protective conductor

S_{PEN} — Cross-sectional area of PEN conductor

S_{KQ} — Short-circuit power of the high voltage network (kVA)

S_{ph} — Cross-sectional area of line conductor (mm^2)

U_0 — Line to neutral nominal voltage of the installation (V)

U_n — Line to line nominal voltage of the installation (V)

X_N — Reactance of the neutral conductor upstream of the circuit being considered

X_{PE} — Reactance of the protective conductor from the main equipotential bonding to the origin of the circuit being considered

X_{PEN} — Reactance of the PEN conductor from the main equipotential bonding to the origin of the circuit being considered (V)

X_Q Reactance upstream of the source

X_T Reactance of the source

X_b Reactance term of mean line-line, line-neutral or line-PE (-PEN) BTS loop impedance

X_{bph} Mean reactance of BTS line conductor, per metre

X_u Reactance of the line conductor upstream of the circuit; $X_u = \sum X_{phase}$

Z_Q Impedance upstream of the source, $Z_Q = \sqrt{R_Q^2 + X_Q^2}$

Z_T Impedance of the source, $Z_T = \sqrt{R_T^2 + X_T^2}$

Z_S $Z_S = Z_Q + Z_T = \sqrt{R_S^2 + X_S^2}$

c Voltage factor

m No load voltage factor

n_N Number of neutral conductors in parallel

n_{PE} Number of protective conductors in parallel

n_{PEN} Number of PEN conductors in parallel

n_{ph} Number of line conductors in parallel

λ Linear reactance of conductors

ρ_0 Resistivity of conductors at 20 °C

ρ_1 Resistivity of conductors at the maximum permissible steady-state operating temperature

ρ_2 Resistivity of conductors at the mean temperature between steady-state temperature and final short-circuit temperature

ρ_3 Resistivity of separate PE conductors at the mean temperature between ambient and final short-circuit temperature.

Equipment data F

▼ **Table F.1**   Values of resistance/metre for copper and aluminium conductors
(*On-Site Guide* Table I1)

Cross-sectional area (mm²)		Resistance/metre or $(R''_1 + R''_2)$/metre (mΩ/m) at 20 °C	
Line conductor	Protective conductor	Copper	Aluminium
1	—	18.10	
1	1	36.20	
1.5	—	12.10	
1.5	1	30.20	
1.5	1.5	24.20	
2.5	—	7.41	
2.5	1	25.51	
2.5	1.5	19.51	
2.5	2.5	14.82	
4	—	4.61	
4	1.5	16.71	
4	2.5	12.02	
4	4	9.22	
6	—	3.08	
6	2.5	10.49	
6	4	7.69	
6	6	6.16	
10	—	1.83	
10	4	6.44	
10	6	4.91	
10	10	3.66	
16	—	1.15	1.91
16	6	4.23	—
16	10	2.98	—
16	16	2.30	3.82
25	—	0.727	1.20
25	10	2.557	—
25	16	1.877	—
25	25	1.454	2.40
35	—	0.524	0.87
35	16	1.674	2.78
35	25	1.251	2.07
35	35	1.048	1.74

Notes:
1 From BS 6360, Table 2.
2 For larger sizes see Tables F.6 and F.7 and Chapter 10.

▼ **Table F.2**   Ambient temperature multipliers C_{F2} to be applied to Table F.1 resistances to convert resistances at 20 °C to other ambient temperatures

Expected ambient temperature (°C)	Correction factor C_{F2}
5	0.94
10	0.96
15	0.98
20	1.00
25	1.02

Note:

The correction factor is given by

$C_{F2} = 1 + 0.004$ {ambient temp – 20 °C}

where 0.004 is the simplified resistance coefficient per °C at 20 °C given by BS EN 60228 for copper and aluminium conductors.

▼ **Table F.3**   Conductor temperature multiplier C_r, to convert conductor resistance at 20 °C to conductor resistance at maximum operating temperature (*On-Site Guide* Table I3)

Conductor Installation	Conductor operating temperature and insulation type		
	70 °C thermoplastic (PVC)	90 °C thermoplastic (PVC)	90 °C thermosetting[4]
Protective conductor not incorporated in the cable and not bunched with the cable [1,3]	1.04	1.04	1.04
Conductor incorporated in a cable or bunched [2,3]	1.20	1.28	1.28

1 See Table 54.2 in Table 8.1 of Chapter 8. These factors apply when protective conductor is not incorporated or bunched with cables, or for bare protective conductors in contact with cable covering. They correct from 20 °C to 30 °C.

2 See Table 54.3 in Table 8.1 of Chapter 8. These factors apply when the protective conductor is a core in a cable or is bunched with cables.

3 The factors are given by
$C_r = 1 + 0.004$ {conductor operating temperature – 20 °C}
where 0.004 is the simplified resistance coefficient per °C at 20 °C given in BS EN 60228 for copper and aluminium conductors.

4 If cable loading is such that the maximum operating temperature is 70 °C, thermoplastic (70 °C) factors are appropriate.

▼ **Table F.4** Coefficients of resistance for conductors

Material	Coefficient of resistance α at 20 °C
Annealed copper	0.00393*
Hard drawn copper	0.00381
Aluminium	0.00403*
Lead	0.00400
Steel	0.0045

* An average value of 0.004 is often used for copper and aluminium.

▼ **Table F.5** Impedance of distribution transformers referred to 415, 480 or 240 V systems as appropriate

Transformer		Resistance per line (Ω)	Reactance per line (Ω)
Type	Rating (kVA)		
Single-phase two-wire in 240 V system	5	0.4300	0.3620
	10	0.1910	0.2060
	15	0.1180	0.1460
	16	0.1080	0.1390
	25	0.0612	0.0944
	25*	0.0570	0.0920
	50	0.0266	0.0496
	50*	0.0270	0.0497
Single-phase three-wire in 240 V system	25	0.0853	0.0943
	50	0.0393	0.0513
	100	0.0165	0.0255
Single-phase three-wire in 480 V system	25	0.2330	0.3650
	50	0.1090	0.1950
	100	0.0445	0.1020
Three-phase in 415 V system and 240 V system	25	0.20800	0.2660
	50	0.08760	0.1440
	100	0.03710	0.0810
	200	0.01580	0.0406
	300	0.00948	0.0281
	315	0.00901	0.0268
	500	0.00509	0.0171
	750	0.00313	0.0115
	800	0.00291	0.0107
	1000	0.00219	0.0086

* Three-wire transformer with links arranged for two-wire output.
From Electricity Association Engineering Recommendation P. 28, Table D6, now available from the Energy Networks Association.

▼ **Table F.6** Thermoplastic (PVC) insulated single-core copper cables, resistance and reactance values at 20 °C

Nominal area of conductors (mm²)	Maximum values of copper conductors at 20 °C	
	Resistance r (mΩ/m)	Reactance x (mΩ/m)
35	0.5240	0.095
50	0.3870	0.094
70	0.2680	0.090
95	0.1930	0.090
120	0.1530	0.085
150	0.1240	0.085
185	0.0991	0.085
240	0.0754	0.083
300	0.0601	0.082

Notes:

1 The values of reactance given above apply only when two cables are installed touching throughout or when three cables are installed in trefoil formation touching throughout.
   Resistance values from BS EN 60228.
   Reactance values from cable manufacturers.

2 For other cable configurations resistance and reactance values for conductors can be derived from Table 4D1B of BS 7671 at 70 °C.
   The resistance component of voltage drop in the three-phase column is divided by ($\sqrt{3}$ x 1.20) and the reactive component divided by $\sqrt{3}$, see Chapter 10.

▼ **Table F.7A** Impedance of conductor and armour for two-core, three-core and four-core thermoplastic (PVC) insulated cables having steel-wire armour

1	2		3		4		5		6		7		8		9	
					Impedance of cable at 20 °C (Ω/km)											
					Steel-wire armour											
					Stranded copper conductors						Solid aluminium conductors					
Nominal cross-sectional area of conductor (mm²)	Copper conductor		Aluminium conductor		Two-core 600/1000 V		Three-core 600/1000 V		Four-core (equal) 600/1000 V		Two-core 600/1000 V		Three-core 600/1000 V		Four-core 600/1000 V	
	r	x	r	x	r	x	r	x	r	x	r	x	r	x	r	x
1.5	12.100	–	–	–	10.20	–	9.5	–	9.50	–	–	–	–	–	–	–
2.5	7.410	–	–	–	8.80	–	8.2	–	7.90	–	–	–	–	–	–	–
4	4.610	–	–	–	7.50	–	7.0	–	4.60	–	–	–	–	–	–	–
6	3.080	–	–	–	6.80	–	4.6	–	4.10	–	–	–	–	–	–	–
10	1.830	–	–	–	3.90	–	3.7	–	3.40	–	–	–	–	–	–	–
16	1.150	0.09	1.910	0.09	3.50	–	3.2	–	2.20	–	3.7	–	3.40	–	2.40	–
25	0.727	0.09	1.200	0.09	2.60	–	2.4	–	2.10	–	2.9	–	2.50	–	2.30	–
35	0.524	0.08	0.868	0.08	2.40	–	2.1	–	1.90	–	2.7	–	2.30	–	2.00	–
50	0.387	0.08	0.641	0.08	2.10	0.3	1.9	0.3	1.30	0.3	2.4	0.3	2.00	0.3	1.40	0.3
70	0.268	0.08	0.443	0.08	1.90	0.3	1.4	0.3	1.20	0.3	2.1	0.3	1.40	0.3	1.30	0.3
95	0.193	0.08	0.320	0.08	1.30	0.3	1.2	0.3	0.98	0.3	1.5	0.3	1.30	0.3	1.10	0.3
120	0.153	0.08	0.253	0.08	1.20	0.3	1.1	0.3	0.71	0.3	–	–	1.20	0.3	0.78	0.3
150	0.124	0.08	0.206	0.08	1.10	0.3	0.74	0.3	0.65	0.3	–	–	0.82	0.3	0.71	0.3
185	0.0991	0.08	0.164	0.08	0.78	0.3	0.68	0.3	0.59	0.3	–	–	0.73	0.3	0.64	0.3
240	0.0754	0.08	0.125	0.08	0.69	0.3	0.60	0.3	0.52	0.3	–	–	0.65	0.3	0.52	0.3
300	0.0601	0.08	0.100	0.08	0.63	0.3	0.54	0.3	0.47	0.3	–	–	0.59	0.3	0.52	0.3
400	0.0470	0.08	–	–	0.56	0.3	0.49	0.3	0.34	0.3	–	–	–	–	–	–

Source for resistances BS 6346, Table H2.

▼ **Table F.7B** Gross cross-sectional area of steel armour wires for two-core, three-core and four-core 600/1000 V thermoplastic (PVC) insulated cables to BS 6346

Nominal area of conductor (mm²)	Cross-sectional area of round armour wires (mm²)						
	Cables with stranded copper conductors				Cables with solid aluminium conductors		
	Two-core	Three-core	Four-core	Four-core (reduced neutral)	Two-core	Three-core	Four-core
1.5*	15	16	17	–	–	–	–
2.5*	17	19	20	–	–	–	–
4*	21	23	35	–	–	–	–
6*	24	36	40	–	–	–	–
10*	41	44	49	–	–	–	–
16*	46	50	72	–	42	46	66
25	60	66	76	76	54	62	70
35	66	74	84	82	58	68	78
50	74	84	122	94	66	78	113
70	84	119	138	135	74	113	128
95	122	138	160	157	109	128	147
120	131	150	220	215	–	138	201
150	144	211	240	235	–	191	220
185	201	230	265	260	–	215	245
240	225	260	299	289	–	240	274
300	250	289	333	323	–	265	304
400	279	319	467	452	–	–	–

* Circular conductors
From BS 6346, Table 36.

▼ **Table F.8A** Impedance of conductor and armour for two-core, three-core and four-core thermosetting (90 °C) insulated cables having steel-wire armour

1	**2**		**3**		**4**		**5**		**6**		**7**		**8**		**9**	
					Impedance in ohms per kilometre of cable at 20 °C (Ω/km)											
					Steel-wire armour											
					Stranded copper conductors						Solid aluminium conductors					
Nominal cross-sectional area of conductor (mm²)	Copper conductor		Aluminium conductor		Two-core 600/1000 V		Three-core 600/1000 V		Four-core (equal) 600/1000 V		Two-core 600/1000 V		Three-core 600/1000 V		Four-core 600/1000 V	
	r	x	r	x	r	x	r	x	r	x	r	x	r	x	r	x
1.5	12.100	–	–	–	10.20	–	9.5	–	8.80	–	–	–	–	–	–	–
2.5	7.410	–	–	–	8.80	–	8.20	–	7.70	–	–	–	–	–	–	–
4	4.610	–	–	–	7.90	–	7.50	–	6.80	–	–	–	–	–	–	–
6	3.080	–	–	–	7.00	–	6.60	–	4.30	–	–	–	–	–	–	–
10	1.830	–	–	–	6.00	–	4.00	–	3.70	–	–	–	–	–	–	–
16	1.150	0.09	1.910	0.09	3.80	–	3.60	–	3.20	–	4.0	–	3.8	–	3.40	–
25	0.727	0.09	1.200	0.09	3.70	–	2.50	–	2.30	–	4.1	–	2.7	–	2.40	–
35	0.524	0.08	0.868	0.08	2.50	–	2.30	–	2.00	–	2.5	–	2.5	–	2.20	–
50	0.387	0.08	0.641	0.08	2.30	0.3	2.00	0.3	1.80	0.3	2.6	0.3	2.2	0.3	1.90	0.3
70	0.268	0.08	0.443	0.08	2.00	0.3	1.80	0.3	1.20	0.3	2.3	0.3	1.9	0.3	1.30	0.3
95	0.193	0.08	0.320	0.08	1.40	0.3	1.30	0.3	1.10	0.3	1.6	0.3	1.4	0.3	1.20	0.3
120	0.153	0.08	0.253	0.08	1.30	0.3	1.20	0.3	0.76	0.3	–	–	–	–	0.82	0.3
150	0.124	0.08	0.206	0.08	1.20	0.3	0.78	0.3	0.68	0.3	–	–	–	–	0.74	0.3
185	0.0991	0.08	0.164	0.08	0.82	0.3	0.71	0.3	0.61	0.3	–	–	–	–	0.67	0.3
240	0.0754	0.08	0.125	0.08	0.73	0.3	0.63	0.3	0.54	0.3	–	–	–	–	0.59	0.3
300	0.0601	0.08	0.100	0.08	0.67	0.3	0.58	0.3	0.49	0.3	–	–	–	–	0.54	0.3
400	0.0470	0.08	–	–	–	–	0.52	–	0.35	–	–	–	–	–	–	–

Source for resistances BS 5467, Table 28.

▼ **Table F.8B** Gross cross-sectional area of steel armour wires for two-core, three-core and four-core 600/1000 V cables with thermosetting insulation to BS 5467

Nominal area of conductor (mm²)	Cross-sectional area of round armour wires (mm²)						
	Cables with stranded copper conductors				Cables with solid aluminium conductors		
	Two-core	Three-core	Four-core	Four-core (reduced neutral)	Two-core	Three-core	Four-core
1.5*	16	17	18	–	–	16	17
2.5*	17	19	20	–	–	17	19
4*	19	21	23	–	–	19	21
6*	22	23	36	–	–	22	23
10*	26	39	43	–	–	26	39
16*	41	44	49	–	40	41	44
25	42	62	70	70	38	42	62
35	62	70	80	76	54	62	70
50	68	78	90	86	60	68	78
70	80	90	131	128	70	80	90
95	113	128	147	144	100	113	128
120	125	141	206	163	–	125	141
150	138	201	230	220	–	138	201
185	191	220	255	250	–	191	220
240	215	250	289	279	–	215	250
300	235	269	319	304	–	235	269
400	265	304	452	343	–	265	304

* Circular conductors
From BS 5467, Table 33.

▼ **Table F.9** Single-core cables having solid aluminium conductors and aluminium strip armour at 20 °C

Nominal area of conductor (mm²)	Impedance values at 20 °C (mΩ/m)			
	Conductor resistance	Conductor reactance	Strip armour resistance	
	r	x	600/1000 V	1900/3300 V
600	0.0515	0.09	Not applicable, not used as earth conductor	

▼ **Table F.10** Cross-sectional areas of steel conduit and trunking

Heavy Gauge Steel Conduit

Nominal diameter (mm)	Minimum steel cross-sectional area (mm²)
16	64.4
20	82.6
25	105.4
32	137.3

Steel Surface Trunking (sample sizes)

Nominal size (mm x mm)	Minimum steel cross-sectional area without lid (mm²)
50 x 50	135
75 x 75	243
100 x 50	216
100 x 100	324
150 x 100	378

Steel Underfloor Trunking (sample sizes)

Nominal size (mm x mm)	Minimum steel cross-sectional area without lid (mm²)
75 x 25	118
100 x 50	142
100 x 100	213
150 x 100	284

▼ **Table F.11** Impedance of steel conduit at 20 °C

Heavy gauge

Nominal conduit size (mm)	Typical impedance [3]	
	Resistance r (mΩ/m) [1]	Reactance x (mΩ/m) [1]
16	3.3	3.3
20	2.4	2.4
25	1.6	1.6
32	1.4	1.4

Light gauge

Nominal conduit size (mm)	Typical impedance [3]	
	Resistance r (mΩ/m) [1]	Reactance x (mΩ/m) [1]
16	4.5	4.7
20	3.7	3.7
25	2.1	2.1
32	1.5	1.5

Notes:

1 Typical values are taken at fault currents greater than 100 A. When I_f is less than 100 A, the tabulated impedances should be doubled.

2 When touch voltages on the conduit are being determined the product of current and resistance (r) gives a good approximation. The inductance of the conduit is reflected into the enclosed cables and is not effective on the conduit itself.

3 The above values are at 20 °C but may be assumed to be independent of temperature and are used for design and verification.

▼ **Table F.12**   Impedance of steel trunking (Note 2)

Size	Typical impedance at 20 °C	
(mm x mm)	Resistance r (mΩ/m)	Reactance x (mΩ/m)
50 x 37	2.96	2.96
50 x 50	2.44	2.44
75 x 50	1.75	1.75
75 x 75	1.37	1.37
100 x 50	1.52	1.52
100 x 75	1.21	1.21
100 x 100	0.87	0.87
150 x 50	1.05	1.05
150 x 75	0.87	0.87
150 x 100	0.81	0.81
150 x 150	0.52	0.52

Notes:

1 When determining touch voltages on the trunking, the product of current and resistance (r) gives a good approximation.

2 The above values are at 20 °C but may be assumed to be independent of temperature and are used for design and verification.

▼ **Table F.13**   Impedance of steel trunking (Note 2)

Size	Typical impedance at 20 °C	
(mm x mm)	Resistance r (mΩ/m)	Reactance x (mΩ/m)
75 x 25	1.28	1.28
75 x 37.5	1.16	1.16
100 x 25	1.08	1.08
100 x 37.5	0.99	0.99
150 x 25	0.74	0.74
150 x 37.5	0.69	0.69
225 x 25	0.52	0.52
225 x 37.5	0.49	0.49

Notes:

1 When determining touch voltages on the trunking, the product of current and resistance (r) gives a good approximation.

2 The above values are at 20 °C but may be assumed to be independent of temperature and are used for design and verification.

▼ **Table F.14** Resistance values (mΩ/m) for 500 V (light duty), multicore, mineral-insulated cables – copper – exposed to touch or thermoplastic covered

Cable ref.	R_1 Conductor resistance 20 °C	R_2 Sheath resistance 20 °C	R_1 Conductor resistance	R_2 Sheath resistance	R_1 Conductor resistance	R_2 Sheath resistance
			Exposed to touch 70 °C sheath		Not exposed to touch 105 °C sheath	
2L1	18.10	3.95	21.87	4.47	24.50	4.84
2L1.5	12.10	3.35	14.62	3.79	16.38	4.10
2L2.5	7.41	2.53	8.95	2.87	10.03	3.10
2L4	4.81	1.96	5.81	2.22	6.51	2.40
3L1	18.10	3.15	21.87	3.57	24.50	3.86
3L1.5	12.10	2.67	14.62	3.02	16.38	3.27
3L2.5	7.41	2.23	8.95	2.53	10.03	2.73
4L1	18.10	2.71	21.87	3.07	24.50	3.32
4L1.5	12.10	2.33	14.62	2.64	16.38	2.85
4L2.5	7.41	1.85	8.95	2.10	10.03	2.27
7L1	18.10	2.06	21.87	2.33	24.50	2.52
7L1.5	12.10	1.78	14.62	2.02	16.38	2.18
7L2.5	7.41	1.36	8.95	1.54	10.03	1.67

Note:
The calculation of $R_1 + R_2$ for mineral-insulated cables differs from other cables, in that the loaded conductor temperature is not tabulated. Table 4G of Appendix 4 of BS 7671 gives normal full load sheath operating temperatures of 70 °C for thermoplastic sheathed types and 105 °C for bare cables not in contact with combustible materials, in a 30 °C ambient. Magnesium oxide is a relatively good thermal conductor, and being in a thin layer, it is found that conductor temperatures are usually only some 3 °C higher than sheath temperatures.

The sheath is of copper to a different material standard to that of the conductors, and the coefficient of resistance 0.004 does not apply. A coefficient of 0.00275 at 20 °C can be used for sheath resistance change calculations. Tables F.14 and F.15 give calculated values of R_1 and R_2 at a standard 20 °C and at standard sheath operating temperatures and these can be used directly for calculations at full load temperatures for devices in Appendix 3 of BS 7671.

▼ **Table F.15** Resistance values (mΩ/m) for 750 V (heavy duty) mineral-insulated cables – copper

Cable ref.	R_1 Conductor resistance 20 °C	R_2 Sheath resistance 20 °C	R_1 Conductor resistance	R_2 Sheath resistance	R_1 Conductor resistance	R_2 Sheath resistance
			Exposed to touch 70 °C sheath		Not exposed to touch 105 °C sheath	
1H10	1.83	2.23	2.21	2.53	2.48	2.73
1H16	1.16	1.81	1.40	2.05	1.57	2.22
1H25	0.727	1.40	0.878	1.59	0.984	1.72
1H35	0.524	1.17	0.633	1.33	0.709	1.43
1H50	0.387	0.959	0.468	1.09	0.524	1.18
1H70	0.268	0.767	0.324	0.869	0.363	0.94

continues

Cable ref.	R₁ Conductor resistance 20 °C	R₂ Sheath resistance 20 °C	R₁ Conductor resistance	R₂ Sheath resistance	R₁ Conductor resistance	R₂ Sheath resistance
			Exposed to touch 70 °C sheath		Not exposed to touch 105 °C sheath	
1H95	0.193	0.646	0.233	0.732	0.261	0.792
1H120	0.153	0.556	0.185	0.63	0.207	0.681
1H150	0.124	0.479	0.15	0.542	0.168	0.587
1H185	0.101	0.412	0.122	0.467	0.137	0.505
1H240	0.0775	0.341	0.0936	0.386	0.105	0.418
2H1.5	12.10	1.90	14.62	2.15	16.38	2.33
2H2.5	7.41	1.63	8.95	1.85	10.03	2
2H4	4.61	1.35	5.57	1.53	6.24	1.65
2H6	3.08	1.13	3.72	1.28	4.17	1.38
2H10	1.83	0.887	2.21	1.005	2.48	1.09
2H16	1.16	0.695	1.40	0.787	1.57	0.852
2H25	0.727	0.546	0.878	0.618	0.984	0.669
3H1.5	12.10	1.75	14.62	1.98	16.38	2.14
3H2.5	7.41	1.47	8.95	1.66	10.03	1.8
3H4	4.61	1.23	5.57	1.39	6.24	1.51
3H6	3.08	1.03	3.72	1.17	4.17	1.26
3H10	1.83	0.783	2.21	0.887	2.48	0.959
3H16	1.16	0.622	1.40	0.704	1.57	0.762
3H25	0.727	0.50	0.878	0.566	0.984	0.613
4H1.5	12.10	1.51	14.62	1.71	16.38	1.85
4H2.5	7.41	1.29	8.95	1.46	10.03	1.58
4H4	4.61	1.04	5.57	1.18	6.24	1.27
4H6	3.08	0.887	3.72	1	4.17	1.09
4H10	1.83	0.69	2.21	0.781	2.48	0.845
4H16	1.16	0.533	1.40	0.604	1.57	0.653
4H25	0.727	0.423	0.878	0.479	0.984	0.518
7H1.5	12.10	1.15	14.62	1.3	16.38	1.41
7H2.5	7.41	0.959	8.95	0.09	10.03	1.18
12H1.5	12.10	0.744	14.62	0.843	16.38	0.912
12H2.5	7.41	0.63	8.95	0.713	10.03	0.772
19H1.5	12.10	0.57	14.62	0.646	16.38	0.698

Note: See the note to Table F.14.

▼ **Table F.16** Resistance values (mΩ/m) for 1000 V (heavy duty) single-core mineral-insulated cables – copper

Number and cross-sectional area of conductors (No. x mm²)	Effective sheath area* (mm²)	Resistance at 20 °C		Cables exposed to touch — Bare and thermoplastic covered cables						Cables NOT exposed to touch — Bare cables					
				Loop resistance at full load (R₁ + R₂)			Loop resistance during earth fault (R₁ + R₂)			Loop resistance at full load (R₁ + R₂)			Loop resistance during earth fault (R₁ + R₂)		
		Con-ductor	Sheath	Single-phase	Three-phase		Single-phase	Three-phase		Single-phase	Three-phase		Single-phase	Three-phase	
				2-wire	3-wire	4-wire	2-wire	3-wire	4-wire	2-wire	3-wire	4-wire	2-wire	3-wire	4-wire
1 x 6	7.8	3.0800	2.20	4.70	4.30	4.10	5.70	5.30	5.10	5.00	4.60	4.40	5.90	5.50	5.20
1 x 10	9.5	1.8300	1.80	3.10	2.80	2.60	3.70	3.40	3.20	3.30	2.90	2.80	3.90	3.50	3.30
1 x 16	12	1.1500	1.50	2.10	1.90	1.70	2.60	2.30	2.10	2.30	2.00	1.80	2.60	2.30	2.20
1 x 25	15	0.7270	1.10	1.50	1.30	1.20	1.80	1.50	1.40	1.60	1.30	1.20	1.80	1.60	1.40
1 x 35	18	0.5240	0.97	1.10	1.00	0.90	1.40	1.20	1.00	1.20	1.00	0.90	1.40	1.20	1.10
1 x 50	22	0.3870	0.79	0.90	0.70	0.64	1.00	0.85	0.77	0.90	0.76	0.68	1.00	0.88	0.79
1 x 70	27	0.2680	0.64	0.70	0.54	0.48	0.75	0.65	0.57	0.70	0.57	0.51	0.77	0.66	0.59
1 x 95	32	0.1930	0.53	0.52	0.42	0.37	0.59	0.51	0.44	0.55	0.44	0.39	0.60	0.52	0.46
1 x 120	37	0.1530	0.46	0.43	0.35	0.30	0.49	0.41	0.36	0.46	0.37	0.32	0.50	0.42	0.37
1 x 150	44	0.1240	0.39	0.36	0.29	0.25	0.40	0.34	0.30	0.38	0.30	0.26	0.42	0.34	0.31
1 x 185	54	0.1010	0.32	0.29	0.23	0.20	0.33	0.27	0.24	0.31	0.25	0.21	0.34	0.28	0.25
1 x 240	70	0.0775	0.25	0.23	0.18	0.16	0.25	0.21	0.19	0.24	0.19	0.17	0.26	0.22	0.19

* The term *effective sheath area* is used because the conductivity of the copper used for the sheath is lower than that used for the conductors. The value shown is calculated as if the materials were the same and enables a direct comparison to be made between conductor and sheath cross-sectional areas.

Notes:

▲ When using single-core cables the protective conductor is made up of: 2 sheaths in parallel for single-phase circuits; 3 sheaths in parallel for three-phase three-wire circuits; and 4 sheaths in parallel for three-phase four-wire circuits.

▲ For cables with line conductors greater than 35 mm² the inductance may need to be considered; however, for single-core cables the inductance will vary with the method of installation and the manufacturer should be contacted for further information.

▼ **Table F.17** Correction factors for change of conductor resistance with temperature. The product of these correction factors and the conductor resistance at 20 °C gives the increase in resistance

Conductor type	Conductor material	Insulation of protective conductor or cable covering					
		70 °C thermoplastic		90 °C thermoplastic		95 °C thermosetting	
		Full load	Fault	Full load	Fault	Full load	Fault
Table 54.2: insulated protective conductor not incorporated in a cable and not bunched with cables, or for separate bare protective conductor in contact with cable covering but not bunched cables.	Copper	0.040	0.300	0.040	0.300	0.040	0.480
	Aluminium	0.040	0.300	0.040	0.300	0.040	0.480
	Steel	0.045	0.338	0.045	0.338	0.045	0.540
	Assumed initial temperature	20 °C	20 °C	20 °C	20 °C	20 °C	20 °C
	Final temperature	30 °C	95 °C	30 °C	95 °C	30 °C	140 °C
Table 54.3: conductor incorporated in a cable or bunched with cables	Copper	0.20	0.38	0.28	0.42	0.28	0.60
	Aluminium	0.20	0.38	0.28	0.42	0.28	0.60
	Assumed initial temperature	20 °C	20 °C	20 °C	20 °C	20 °C	20 °C
	Final temperature	70 °C	115 °C	99 °C	125 °C	90 °C	170 °C
Table 54.4: protective conductor as a sheath or armour of a cable	Aluminium	0.16	0.44	0.24	0.48	0.24	0.48
	Steel	0.18	0.50	0.27	0.54	0.27	0.54
	Lead	0.16	0.44	0.24	0.48	0.24	0.48
	Assumed initial temperature	20 °C	20 °C	20 °C	20 °C	20 °C	20 °C
	Final temperature	60 °C	130 °C	80 °C	140 °C	80 °C	140 °C

▼ **Table F.18A** 230 V single-phase supplies up to 100 A (from Engineering Recommendation P25): estimated maximum prospective short-circuit current at the electricity distributor's cut-out based on declared level of 16 kA (0.55 p.f.) at the point of connection of the service line to the LV distributing main

Length of service line (m)	Up to 25 mm² Al or 16 mm² Cu service cable or overhead line		35 mm² Al or 25 mm² Cu service cable or overhead line (looped service)	
	p.s.c.c. (kA)	p.f.	p.s.c.c. (kA)	p.f.
0	16.0	0.55	16.0	0.55
1	14.8	0.63	15.1	0.61
2	13.7	0.69	14.3	0.66
3	12.6	0.74	13.5	0.70
4	11.7	0.78	12.7	0.74
5	10.8	0.82	12.0	0.77
6	10.1	0.84	11.4	0.79
7	9.4	0.86	10.8	0.82
8	8.8	0.88	10.3	0.83
9	8.3	0.89	9.7	0.85
10	7.8	0.91	9.3	0.86
11	7.4	0.92	8.8	0.88
12	7.0	0.92	8.4	0.89
13	6.6	0.93	8.1	0.90
14	6.3	0.94	7.7	0.91
15	6.0	0.94	7.4	0.91
16	5.7	0.95	7.1	0.92
17	5.5	0.95	6.9	0.92
18	5.3	0.96	6.6	0.93
19	5.1	0.96	6.4	0.93
20	4.9	0.96	6.2	0.94
21	4.7	0.96	6.0	0.94
22	4.5	0.97	5.8	0.95
23	4.4	0.97	5.6	0.95
24	4.2	0.97	5.4	0.95
25	4.1	0.97	5.3	0.95
26	3.9	0.97	5.1	0.96
27	3.8	0.98	5.0	0.96
28	3.7	0.98	4.8	0.96
29	3.6	0.98	4.7	0.96
30	3.5	0.98	4.6	0.96
35	3.1	0.98	4.0	0.97

continues

▼ **Table F.18A** *continued*

Length of service line (m)	Up to 25 mm² Al or 16 mm² Cu service cable or overhead line		35 mm² Al or 25 mm² Cu service cable or overhead line (looped service)	
	p.s.c.c. (kA)	p.f.	p.s.c.c. (kA)	p.f.
40	2.7	0.99	3.6	0.98
45	2.5	0.99	3.3	0.98
50	2.2	0.99	3.0	0.98

▼ **Table F.18B** 415 V three-phase supplies, 25 kA p.s.c.c. (from Engineering Recommendation P26): estimated maximum p.s.c.c. at the electricity distributor's cut-out based on declared levels of 25 kA (0.23 p.f.) at the point of connection of the service line to the busbars in the distribution substation

Length service line (m)	Service line cross-sectional area									
	95 mm² Al		120 mm² Al		185 mm² Al		240 mm² Al		300 mm² Al	
	p.s.c.c. (kA)	p.f.	p.s.c.c. (kA)	p.f.	p.s.c.c. (kA)	p.f.	p.s.c.c. (kA)	p.f.	p.s.c.c. (kA)	p.f.
0	25.0	0.2	25.0	0.2	25.0	0.2	25.0	0.2	25.0	0.2
5	23.1	0.4	23.3	0.3	23.7	0.3	23.8	0.3	23.9	0.3
10	21.1	0.5	21.7	0.4	22.4	0.4	22.7	0.3	22.8	0.3
15	19.2	0.6	20.1	0.5	21.1	0.4	21.6	0.4	21.8	0.3
20	17.5	0.6	18.6	0.6	20.0	0.5	20.5	0.4	20.9	0.4
25	16.0	0.7	17.2	0.6	18.9	0.5	19.6	0.4	20.0	0.4
30	14.6	0.7	16.0	0.7	17.9	0.5	18.7	0.5	19.2	0.4
35	13.5	0.8	14.9	0.7	16.9	0.6	17.8	0.5	18.4	0.4
40	12.4	0.8	13.9	0.7	16.1	0.6	17.1	0.5	17.7	0.5
45	11.6	0.8	13.0	0.7	15.3	0.6	16.3	0.5	17.1	0.5
50	10.8	0.8	12.3	0.8	14.6	0.6	15.7	0.6	16.4	0.5

▼ **Table F.18C** 415 V three-phase supplies, 18 kA p.s.c.c. (from Engineering Recommendation P26): estimated maximum p.s.c.c. at the electricity distributor's cut-out based on declared levels of 18 kA (0.5 p.f.) at the point of connection of the service line to the LV distribution main

| Length service line (m) | Service line cross-sectional area | | | | | | | | | | | | | |
| | Up to 35 mm² Al | | 70 mm² Al | | 95 mm² Al | | 120 mm² Al | | 185 mm² Al | | 240 mm² Al | | 300 mm² Al | |
	p.s.c.c. (kA)	p.f.	p.s.c.c. (kA)	p.f.	p.s.c.c. (kA)	p.f.	p.s.c.c. (kA)	p.f.	p.s.c.c. (kA)	p.f.	p.s.c.c. (kA)	p.f.	p.s.c.c. (kA)	p.f.
0	18.0	0.5	18.0	0.5	18.0	0.5	18.0	0.5	18.0	0.5	18.0	0.5	18.0	0.5
5	14.8	0.7	16.2	0.6	16.6	0.6	16.8	0.6	17.1	0.5	17.2	0.5	17.3	0.5
10	12.2	0.8	14.5	0.7	15.3	0.6	15.7	0.6	16.2	0.6	16.5	0.5	16.6	0.5
15	10.2	0.8	13.1	0.7	14.1	0.7	14.7	0.6	15.4	0.6	15.8	0.6	16.0	0.6
20	8.8	0.9	11.9	0.8	13.0	0.7	13.7	0.7	14.7	0.6	15.1	0.6	15.4	0.6
25	7.6	0.9	10.8	0.8	12.1	0.7	12.9	0.7	14.0	0.6	14.6	0.6	14.9	0.6
30	6.8	0.9	9.9	0.8	11.3	0.8	12.2	0.7	13.4	0.7	14.0	0.6	14.4	0.6
35	6.0	0.9	9.1	0.9	10.6	0.8	11.5	0.7	12.9	0.7	13.5	0.6	13.9	0.6
40	5.5	0.9	8.5	0.9	9.9	0.8	10.9	0.8	12.3	0.7	13.0	0.6	13.5	0.6
45	5.0	0.9	7.9	0.9	9.3	0.8	10.3	0.8	11.8	0.7	12.5	0.6	13.1	0.6
50	4.6	0.9	7.4	0.9	8.8	0.8	9.8	0.8	11.4	0.7	12.1	0.7	12.7	0.6

▼ **Table F.19** Gates for specified pre-arcing times for BS EN 60269-2 and BS 88 'gG' and 'gM' fuse-links (from Table 3 of BS EN 60269-1)

1	2	3	4	5	6	7	8	9
I_n for 'gG' I_{ch} for 'gM'[1]	I_{min} (10 s)[2]	I_{max} (5 s)[3]	I_{min} (0.1 s)	I_{max} (0.1 s)	I (0.4 s) Fig. 3A3 BS 7671	Max Z_s for 0.4 s disconnection Table 41.2	I (5 s) Fig. 3A3 BS 7671	Max Z_s for 5 s disconnection Table 41.4
(A)	(A)	(A)	(A)	(A)	(A)		(A)	
16	33	65	85	150	90	2.56	55	4.18
20	42	85	110	200	130	1.77	78	2.95
25	52	110	150	260	170	1.35	100	2.3
32	75	150	200	350	220	1.04	125	1.84
40	95	190	260	450	290		170	1.35
50	125	250	350	610	380		220	1.04
63	160	320	450	820	500		280	0.82
80	215	425	610	1100	750		400	0.57
100	290	580	820	1450	1000		520	0.46
125	355	715	1100	1910	1250		680	0.34
160	460	950	1450	2590	1600		820	0.28
200	610	1250	1910	3420	2200		1200	0.19
250	750	1650	2590	4500				
315	1050	2200	3420	6000				
400	1420	2840	4500	8060				
500	1780	3800	6000	10600				
630	2200	5100	8060	14140				
800	3060	7000	10600	19000				
1000	4000	9500	14140	24000				
1250	5000	13000	19000	35000				

Notes to Table F.19:

1 Breaking range and utilization category

The first letter indicates the breaking range:
- 'g' fuse-links (full-range breaking-capacity fuse-link);
- 'a' fuse-links (partial-range breaking-capacity fuse-link).

The second letter indicates the utilization category:
- 'gG' indicates fuse-links with a full-range breaking capacity for general application;
- 'gM' indicates fuse-links with a full-range breaking capacity for the protection of motor circuits;
- 'aM' indicates fuse-links with a partial range breaking capacity for the protection of motor circuits;
- 'gD' indicates time-delay fuse-links with a full-range breaking capacity;
- 'gN' indicates non-time-delay fuse-links with a full-range breaking capacity.

2 I_{min} (10 s) is the minimum value of current for which the pre-arcing time is not less than 10 s.

3 I_{max} (5 s) is the maximum value of current for which the operating time is not more than 5 s.

4 'gG' fuse-links can be used for the protection of motor circuits, if their characteristics are such that the link is capable of withstanding the motor starting current.

Index

Index

Index